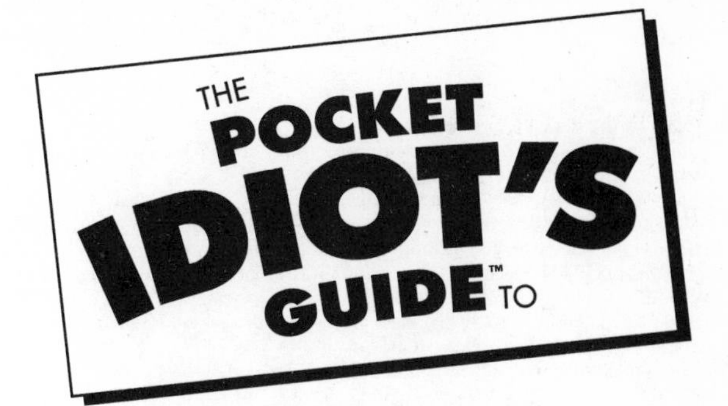

Patents

by Gregory Scott Smith and
Robert J. Frohwein

ALPHA
A member of Penguin Group (USA) Inc.

ALPHA BOOKS

Published by the Penguin Group

Penguin Group (USA) Inc., 375 Hudson Street, New York, New York 10014, U.S.A.

Penguin Group (Canada), 10 Alcorn Avenue, Toronto, Ontario, Canada M4V 3B2 (a division of Pearson Penguin Canada Inc.)

Penguin Books Ltd, 80 Strand, London WC2R 0RL, England

Penguin Ireland, 25 St Stephen's Green, Dublin 2, Ireland (a division of Penguin Books Ltd)

Penguin Group (Australia), 250 Camberwell Road, Camberwell, Victoria 3124, Australia (a division of Pearson Australia Group Pty Ltd)

Penguin Books India Pvt Ltd, 11 Community Centre, Panchsheel Park, New Delhi - 110 017, India

Penguin Group (NZ), Cnr Airborne and Rosedale Roads, Albany, Auckland, New Zealand (a division of Pearson New Zealand Ltd)

Penguin Books (South Africa) (Pty) Ltd, 24 Sturdee Avenue, Rosebank, Johannesburg 2196, South Africa

Penguin Books Ltd, Registered Offices: 80 Strand, London WC2R 0RL, England

International Standard Book Number: 1-59257-229-4
Library of Congress Catalog Card Number: 2004111429

06 05 04 8 7 6 5 4 3 2 1

Interpretation of the printing code: The rightmost number of the first series of numbers is the year of the book's printing; the rightmost number of the second series of numbers is the number of the book's printing. For example, a printing code of 04-1 shows that the first printing occurred in 2004.

Printed in the United States of America

Contents

Appendixes

Introduction

Owning a home used to be considered the American dream. However, this dream has been replaced by many people—super-sized, if you will. For many, the American dream has shifted to starting and owning your own business. But how do you get started? One way is with an idea.

Should you tell anyone your idea? Yes—but first you must figure out a way to protect the idea and yourself.

Fortunately, the legal system has developed a way you can protect your idea. Lawmakers realized that without adequate protection of ideas, people would not have the incentive to create and release inventions. In this book we'll introduce you to the process of protecting ideas, a process known as *patenting*.

Extras to Help You Along

This book also has useful information provided in sidebars throughout the text.

Just the Facts

This box introduces fascinating stories about patents.

Legal-Ease

These sidebars condense legalese into easy-to-understand language.

Objections

Mind these cautionary words if you want to stay inside the boundaries of the law.

Good Counsel

This box provides simple tips to help you recognize and protect your ideas.

Acknowledgments

We'd like to thank Paul Dinas, Michael Koch, and Megan Douglass for their help in putting this book together.

In memory of Gene S. Zimmer who inspired me to cross into the world of patent attorneys and to all my friends from Jones & Askew—Greg

To my wife Dana, and my children, Michael and Lizabeth, who put up with me as I spent my nights and weekends writing this book—Rob

Chapter 1

Patent Basics

In This Chapter

- Recognizing an invention
- Seeing yourself as an inventor
- Is your invention patentable?
- Don't lose your rights
- How patents differ from copyrights and trademarks

You have probably heard the words "patent pending" many times before. But what does that mean? Can you slap that phrase on your latest middle-of-the-night idea and be assured that no one else will steal it? Not exactly.

The U.S. Constitution provided the United States Congress with the authority to pass laws that, for a limited time, protect inventions to promote the progress of science and useful arts. This basic concept has turned into the U.S. patent system. Since the writing of that famous passage in the Constitution, more than 6.5 million patents have issued.

In this chapter, we're going to introduce you to patent basics: what they are and what they're not. We're also going to help you recognize that you can be considered an inventor or the owner of a U.S. patent. We're going to remove some of the mystery around this topic and help you start taking the steps necessary to pursue patent protection.

What Is an Invention?

The first and most difficult question we need to confront is what constitutes an invention that can be patented. Most people, understandably, think that you need to come up with the cure for cancer to have something worth patenting. This is simply not true. Some of the most worthwhile inventions come in the form of small improvements to existing products.

A nice example is the red squiggly line that appears under misspelled words on your word processor (a quite relevant example as we're typing up this book!). Some time ago, an individual came up with the word processing program and all of the initial important features in the program. They did think of spell check pretty early on. The spell-check feature enables you to review an entire document or portion of text at your command. However, at that time, they didn't think of the red squiggly line. Only after years of misspelled words and the forgetfulness of folks to run the spell-check program did someone say, "How about we let people know the moment they misspell a word?" And voilà, the idea for the red squiggly line was born.

Just the Facts

On July 31, 1790, Samuel Hopkins was issued the first patent for a process of making potash, an ingredient used in fertilizer. The patent was signed by President George Washington.

Seeing Yourself as an Inventor

You may be saying to yourself right now, "I'm not an inventor! The things that I think up don't have any value." Did you know that most of the great inventors out there initially echoed these same words? The reality is that a large percentage of inventions go legally unprotected because inventors discount the uniqueness and value of their creations. We highly recommend that you consult with a professional before you determine that your invention is not so special.

Protecting Your Invention

Great, so now you realize you've got an invention. Most people with a good idea have a natural fear of it being stolen by some unscrupulous individual or company. The only way to definitely avoid having this happen is to protect the idea. You can protect your idea through a variety of techniques, including the following:

- **Keep your mouth shut!** One easy way is to avoid telling anyone your idea. The reality is that if you ever want to make money off of your idea, you're going to have to tell someone! That being said, we've known many people who were so fearful of sharing an idea that they never did, and then someone else came out with the idea. That's not a good plan!

- **Sign a nondisclosure agreement.** A very popular approach to disclosing an idea is having the party that you're disclosing to sign a *nondisclosure agreement* or *NDA* for short. An NDA will require the other party not to disclose your idea to anybody else.
- **Get a patent!** That's what we're writing about in this book. The best overall protection you can get for your idea is an issued patent. Also, when you file your patent, your need to keep hush-hush all the time decreases, as you've established that you were the first to come up with the idea (unless someone filed a patent on the same idea before you).

Patenting Your Idea

In today's high-tech world, patent protection and the assertion of patent rights has become commonplace for both individuals and corporations. Today we are crying out that you need to "circle the wagons" to provide protection for your inventions and for the value that can be obtained through your inventions. This protection is obtained through applying for and obtaining a patent. In fact, pursuit of patents has increased dramatically over the last decade.

A *patent* is a grant by the government that allows an inventor to prevent others from doing certain things that require the use of the invention. Specifically, the inventor can prevent others from making, using, selling, offering to sell, or importing products or services that incorporate or require the invention. Essentially, the grant gives the inventor a monopoly in the invention for a limited period of time (while the patent is in force).

What If There Were No Patent System?

Picture this: A young chemist slaves for 10 years at great financial risk but finally discovers the vaccine for the horrible plague of the year 2100. She produces the first batch of vaccines and distributes it to a group of needy individuals. One of these individuals grabs the vaccine, figures out the recipe, and starts producing it himself. Worse yet, the thief claims that he came up with the vaccine first! Suddenly, what started out as a creative pursuit in the face of significant financial risk has ended in a flurry of lost opportunities and broken dreams. Was all the hard work worth it? Will the next young chemist who comes face-to-face with a horrible plague be willing to dedicate 10 precious years of his or her life to identify a vaccine? Probably not.

Providing the young chemist with the ability to protect her invention—the vaccine—is a vital part of encouraging folks like her to take risk and achieve great things. With legal protection, she knows that she can obtain the recognition and the value that she has created without paranoia that someone else will steal her life's work. Without this confidence, people may become reluctant to personally sacrifice to better our society. Thanks to patent protection, the chemist can prevent others from taking advantage of her invention without her permission.

Even better, with a patent, the chemist can sell this new vaccine to a huge pharmaceutical company with the confidence that if anyone else tries to copy and sell the vaccine, she can stop them. With the money earned, the young chemist can open her own chemical research laboratory, retire to the ski slopes of Colorado, or simply relax knowing that she is set for life. Now, when the next horrible plague comes along, there will be hordes of young chemists working late into the night to invent a new vaccine and earn their chance at fame and fortune.

Actions That Could Ruin Your Chances

We've generally talked about the potential need to patent your idea. However, not only is someone stealing your idea a risk, but there are also a few mistakes you can make that may ruin your chances of ever obtaining a patent. We'll briefly go into them in the following list:

- **Failing to apply for a patent within one year.** You have exactly one year from the date of either publicly disclosing your invention or offering a product that incorporates your invention to file a U.S. patent application. This can be either a provisional or nonprovisional application (see Chapter 2 for more details). If you do not file the application within the time period, you have *dedicated* your invention to the public, which means anyone can use your invention without your ability to stop them. If you're thinking about obtaining any foreign patent rights, you have to file before you publicly disclose your invention. This means that the foreign rules are even stricter than the U.S. rules!

- **Disclosing your invention publicly.** So what does it mean to publicly disclose your invention? You should not disclose your invention to a person in the absence of signing a nondisclosure agreement with that person. Oftentimes, companies disclose new inventions at trade shows or during press interviews. These are classic examples of public disclosure.
- **Using invention-assistance companies.** You may have seen advertisements on TV from companies asking you to submit your ideas to them so that they may help you make money off of them. The sad reality is that most of these companies are not helpful to the average inventor. These companies charge large sums of money to unsuspecting investors. Of course, there are probably a few companies in this line of work that are

legitimate, but you should proceed very carefully before you invest time and money in one of these companies. Most important, you should ask many questions (and get answers in writing) regarding the percent of their customers who actually generate more income than expense. Further, you should ask for and check a variety of references.

The Date of Invention

As you will learn as you read on, it is important to know the date of your invention. In the United States, patent rights are granted on the basis of the first to invent. Thus, it is very important to be able to prove the date of your invention. In general, it is a good practice to keep written notes in a bound notebook with page numbers; to sign and date each entry in the notebook; and, if possible have a third party read, sign, and date each page. These are ideal conditions for establishing a date of invention.

The date of invention is the first date that you reduce your invention to practice. The date of invention is not simply the date that you conceive of the idea—you must first reduce the invention to practice. Does this mean that you have to create a working model? No, reduction to practice is simply placing the invention in a complete form, which can include a working model or also a written description or drawing of the invention.

If you cannot establish evidence as to the date of your invention, preparing and filing a patent application qualifies as a constructive reduction to practice and as your date of invention.

The Requirements for Getting a Patent

But can anyone get a patent on any idea they create? No, to obtain patent protection for an invention, the invention must fulfill the following three requirements; it must be …

- Useful
- Novel or new
- Nonobvious

If the invention meets these requirements for being protected by a patent and the patent application is otherwise in order, then the United States Patent and Trademark Office (USPTO) will issue you a patent. The exact meanings of the terms useful, novel, and nonobvious were initially defined legally and have been interpreted extensively over the years by the members of the USPTO and courts of law.

Useful

The easiest hurdle to cross is the useful requirement. The USPTO will grant a U.S. patent to whomever invents or discovers any new and useful process, machine, manufacture, or composition of matter, or any new and useful improvement thereof, as long as such invention is also novel and nonobvious. What does the USPTO consider useful? Well, a quick search of the World Wide Web using the term "wacky patent" will quickly show you that this threshold is quite low. For instance, the USPTO considered the following inventions useful and qualified for patenting:

- Vanity fish bait
- A locket for storing your chewing gum
- A device for creating dimples
- Protective eyewear for chickens

As a general rule of thumb, as long as your invention falls in the realm of patentable subject matter (this topic is more fully addressed in Chapter 2), it is considered useful.

Novel or New

The novel and nonobvious requirements are a bit more difficult to meet. To be novel, you must be the first inventor. You cannot meet the novelty requirement if it is shown that someone else knew about, used, or invented your invention prior to you. The USPTO will agree that your invention meets the novel requirement if your invention does not meet the following criteria:

- Your invention was known or used by others in the United States, or patented or described in a printed publication in the United States or a foreign country, before the date of your invention.

- More than one year before the date of filing a patent application for your invention, your invention was patented or described in a printed publication in the United States or a foreign country or was in public use or on sale in the United States.
- Your invention is considered abandoned because you have allowed a long period of time to lapse without furthering development of your invention or filing a patent application.
- You were not the actual inventor of the subject matter sought to be patented.

To show your invention is not novel, the USPTO has to find public documents that describe each of the elements of the invention you are claiming and that satisfy the previous list of criteria. Typically, the USPTO uses published patents, trade journal articles, or similar publications.

Nonobvious

If the USPTO cannot find a single document that describes each of the elements of your invention, then your invention satisfies the novel requirement.

To be nonobvious, your invention must not be an obvious extension of what is already available. To satisfy the obviousness requirement, the USPTO must determine that your invention would not have been obvious to one skilled in the field of your invention.

If the USPTO cannot find a single document that describes your invention but can find multiple documents that together describe your invention, they may allege that your invention is obvious based on the combination of the documents. In addition, the USPTO may allege that your invention is obvious if they identify one or more documents that disclose several elements of your invention and believe that the missing elements would be obvious additions.

Applying for and obtaining a patent can be a very complex and lengthy process, but don't be discouraged. Consider a young boy who would complain about what a pain it was to mend the fence around his parent's farm. In response to his complaints, his father would always reply, "Yeah, but not quite as painful as waking up every morning and having to gather up your cows." Although complex, by following step-by-step instructions, the job can be accomplished. And even more important, if you give up, you may forfeit valuable rights and lose your chance at fame and fortune!

They're Not Copyrights or Trademarks!

Many people lump together patents, copyrights, and trademarks. Although they all have one thing in common (they're all forms of intellectual property protection), they are not the same and should not be used interchangeably. Each is a form of legal protection covering different items created by your mind.

Copyrights

Copyrights protect the expression of an idea and are usually referred to as *works*. These works fall into a variety of categories, including literary works (books), sound recordings (music), visual arts (statues), performing arts (movies), and serials and periodicals (newspapers).

Copyrights provide the owner with certain exclusive rights such as the right to reproduce the work, distribute it, perform it, make a derivative work (for example, the movie version of a book), and display it.

Trademarks

Trademarks protect those words and symbols (called *marks*) that help consumers identify the origin of the goods. These marks typically act as the *brands* a company may use to offer its products or services. For example, the name *Coke* identifies a drink sold by the Coca-Cola Company. Trademarks are also associated with a defined set of products or services, or both. Therefore, a mark that covers the distribution of sink faucets may be the same as a mark that covers airline services (for example, Delta).

Trademark protection allows the trademark owner to prohibit others from using a mark that is confusingly similar. If a person or company is using a mark that is too similar to a registered trademark, that mark may be found to be infringing the registered mark.

Good Counsel

Your intellectual property strategy should seek to protect your company through a combination of intellectual properties—patents, trademarks, copyrights, and trade secrets.

The Least You Need to Know

- Patents are used to protect inventions or ideas such as processes, systems, and methods for accomplishing something of value.
- Your patent is protected by federal law after issuance.
- Generally, patents protect inventions; copyrights protect expressive works; and trademarks protect names, symbols, brands, and other marks that are distinctive to your good or service.

Chapter 2

Should I File a Patent Application?

In This Chapter

- Recognizing an invention
- Seeing yourself as an inventor
- Don't lose your rights
- Differences between copyrights and trademarks

Now that you know some of the benefits of the patent system, take a step back and understand the steps you can go through to make sure you're ready to undertake what is a very time-intensive and prolonged effort—the patent process.

In this chapter, we'll help you understand the different types of patents that are recognized by the patent office. We'll also explain some pre-application steps you can take to ensure that you can obtain a patent and whether you even really want to obtain a patent.

Patentable Subject Matter

The first thing you need to consider when contemplating pursuing a patent is whether your invention would be considered *patentable subject matter*. Federal laws specifically outline the types of ideas that may be subject to patent protection. These items include a process, machine, article of manufacture, or composition of matter (or any improvements on any of these).

Types of Patents

The patent office grants three types of patents: utility, design, and plant patents.

Utility Patents

These patents are issued for the invention of a new and useful process, machine, manufacture, or composition of matter or an improvement to any one of these items. We'll go through each in turn:

- **Method and process patents.** Process patents cover the series of actions, changes, or functions bringing about a desired result or solution to a challenge. Some of the most popular process-oriented patents include …
    - **Business method patent.** A business method is a unique, novel, and useful approach to pursuing a particular approach to a business challenge.

A classic example of a business method is the Amazon.com *one-click* ordering system, which enables customers to quickly order items from Amazon.com. For many years, business methods were rejected as inappropriate for patent protection. However, since the late 1990s, business methods have been accepted as patentable subject matter, and the number of business method patent applications has mushroomed.

- **Software patents.** Until the early 1990s, it was difficult to obtain a patent on a software process, as the patent office and courts considered software to be an expression of a mathematical formula and, thus, not patentable. However, now software is generally accepted as appropriate for patent protection.

- **Articles of manufacture patents.** These items would include any item that is man-made. These could include tools, products, medical devices … the list goes on and on.
- **Machine patents.** A machine is a mechanical device with a multitude of parts that work together to produce a specific result. Included in the definition of a machine are manually operated and automatic devices.

Some machine patent applications cover an entire device (a car), and sometimes they just cover a subcomponent (a carburetor).

- **Compositions of matter patents.** Compositions of matter are a combination of naturally occurring substances that have properties different from their individual ingredients. The best example of a composition of matter is a medicine, detergent, or chemical.

Design Patents

Design patents are focused on the visual and decorative aspects contained in or on an article of manufacture. A design patent only protects the appearance of an invention, not the usefulness or structural aspects of it. The types of items that are typically protected under a design patent are as follows:

- Bridges
- Buildings
- The look of a toothbrush
- The designs of various basketball sneakers
- The Rolls Royce hood ornament

Plant Patents

These patents relate to plants (not manufacturing plants, but the leaf variety!). You can only patent plants that you've invented or discovered and asexually reproduced a distinct and new variety of plant from. There are a few exceptions to this rule. If your invention falls into this category, we recommend that you visit the USPTO website at www.uspto.gov/web/offices/pac/plant/index.html.

You Can't Patent That!

For the most part, items that would be subject to either copyright protection or trademark protection cannot be patented. A small exception to this rule would be the ability to both trademark and patent a particular design. Other items that are not patent-eligible include the following:

- **Laws of nature and natural phenomena.** Essentially, items that are natural laws, such as the law of gravity or $E=mC^2$, are not patentable.
- **Products of nature.** New types of emeralds or jewels discovered in a mine cannot be patented.
- **Mathematical formulas.** These are similar to natural principles and, therefore, not patentable.

Your Rights as a Patent Owner

As a patent owner, you do not have the affirmative right to produce your invention. Instead, you have the right to stop others from making, using, selling, or importing your patented invention. In fact, you'll have the right to stop others for 20 years after you first filed your application (there are a few exceptions to this term rule covered in Chapter 8). So what's the difference between an affirmative right to produce your invention versus the ability to stop others from doing so? It may actually be the case that you're not even entitled to produce your invention. Typically, this happens when you've made an improvement to someone else's invention but have not acquired the right to this underlying invention. Assuming this is *not* the case, you will enjoy a long-term monopoly to your invention.

Patent Infringement

So what happens when someone makes, uses, sells, or imports your patented invention without your permission? They're infringing on your patent! If and when this happens, you can take a variety of actions, including the following:

- **Send them a cease-and-desist letter.** It's quite possible that the infringing party is not aware of your exclusive right. If this is the case, sometimes an initial letter informing them of your rights and asking them to refrain from infringing on your patent may do the trick.
- **File a patent infringement suit.** Filing suit is a very expensive approach—many patent infringement suits now cost between $1 and $4 million to pursue. Therefore, it's probably best to attempt to work out any issues amicably before resorting to a lawsuit.

- **Wait and do nothing.** On occasion, it is worthwhile not to pursue either a cease-and-desist or a litigation strategy for the short term. You may want a competitor to become overly reliant on your patented product line before informing them that they need to stop (or requesting them to pay a license fee). Stopping a fledgling business after their first 10 sales won't exactly create a lot of financial benefit on your side.

> **Good Counsel**
>
> Make sure you're fairly confident of infringement prior to sending a cease-and-desist letter. Inappropriate accusation of patent infringement can be detrimental to the continued validity of your patent and even lead to the other party filing a lawsuit and asserting that their products do not infringe your patent.

The Prior Art Search and Patentability Opinions

So now you've recognized your invention, determined that it is patentable subject matter, and even categorized it as, for example, a utility patent. Should you move forward with a patent application? There are many things you may still need to consider prior to pursuing a patent application, including whether to obtain a prior art search and related patentability opinion.

A prior art search seeks to uncover information, including existing patents (both domestic and foreign) and literature (like white papers, newspapers, or magazine articles), that disclose the invention you're claiming in your application. As you'll learn, the patent office conducts a similar search of your invention while they're reviewing your application. To the extent that someone has come up with your invention or parts of it first, the patent office will limit or eliminate your ability to obtain protection. Therefore, a prior art search will help you understand ahead of time what's likely to happen. It may be worthwhile to save money on a patent application that is unlikely to lead to an issued patent.

After you receive the prior art results, it may be difficult to project exactly how much protection you'll be able to get from the patent office. As a result, it may be worthwhile to have a patent attorney examine the results of the prior art search and provide you with a more detailed explanation of what you're likely to obtain in the way of protection. This is called a *patentability opinion*.

Good Counsel

Please note that if you do conduct a prior art search and still apply for a patent, you will be required to disclose the relevant results of your prior art search to the patent office. This falls under your information disclosure requirements.

Other Considerations Before You File

The following list outlines additional items to consider before you file a patent application:

- **Cost.** Can you afford the patent process? The patent process can be quite expensive, ranging from a few thousand dollars to more than $10,000. You'll learn in Chapter 6 about provisional applications, which may greatly reduce your initial outlay of funds. However, provisional applications will ultimately need to be converted to a non-provisional application and the cost associated with that will remain significant. Make sure you're prepared to invest the money before you expend sums on prior art searches and patentability opinions.
- **Business reality.** Did you know that the median value of a patent is zero and the average value of a patent is well in the millions? Do you know why? Because most people file patent applications and then never pursue their business ideas. That means that thousands of dollars are wasted. Although you shouldn't take forever to file a patent application (because someone might file before you), you should certainly understand whether you have the energy and capability to pursue the intended business. Perhaps the business idea is simply to file

the patent, get it granted, and then either license or sell the concept. If that's the case, fine. But if your intention is to be the person building the business, make sure you are absolutely ready before you plunk down thousands on an application.

- **Time.** Realize that you won't receive a patent for a very long time. As you'll learn in Chapter 7, the timeline for receiving an issued patent is about three years and perhaps even longer. Therefore, the patent process requires patience. If you don't have patience, don't move forward.
- **Expectations.** Most patent applications face initial rejections and the ultimate breadth of the patent likely won't be what is claimed in the original application. You need to adjust your expectations for this reality and make sure you can still achieve your objectives if the patent issued is ultimately narrower than you were hoping for.

The Least You Need to Know

- There are three types of patents: utility, design, and plant patents.
- As a patent owner, you have the right to exclude others from making, using, selling, or importing items that infringe on your patent.
- You should conduct a prior art search and get a patentability opinion to ensure that the patent you would like to pursue will be valuable.

Chapter 3

The Conventional Patent Application

In This Chapter

- The parts of a patent application
- What to include in the specification
- How to construct your claims
- When to call an attorney

An interesting thing about patents is that there are only two classes of people who can prepare, file, and prosecute a patent application with the USPTO. The first class of people includes patent agents and patent attorneys. Patent agents are individuals who have the required necessary technical background to take the patent bar test and have successfully passed the test. Patent attorneys are individuals who have met the requirements for being a patent agent but who have also obtained a legal degree and passed a state-administered bar exam.

The other class of people are inventors! So are we saying that any person with absolutely no formal training can prepare, file, and prosecute a patent application with the U.S. Patent Office? Yes, as long as that person is named as a valid inventor on the patent application, he or she can do the work of a patent attorney—without all the years of schooling!

Does this mean that the skills possessed by a patent attorney are unnecessary? Certainly not. A patent application is a very technical and complex document that includes several significantly different parts. It is very important to make sure that the patent is written correctly the first time because there are really no second chances. One subtle word in a claim can drastically alter the scope and strength of that claim. One statement in the detailed description can greatly limit the claim coverage. It is important to know what the issues are, the pitfalls, and the potential mistakes when preparing, filing, and prosecuting a patent application. In addition, prosecuting a patent application before the USPTO is a very complex process. Failing to meet certain deadlines can result in an abandonment of the patent application and a forfeiture of your rights.

A well-seasoned patent attorney is keenly aware of the snares and pitfalls that can easily limit the scope of protection obtained through a patent. An inventor without any formal training may be able to muddle through the process to obtain a patent, but there is a possibility that the patent may be greatly limited or weak. One of our goals in writing this book is to give you instructions and understanding to help bridge the gap between a seasoned patent attorney and an inventor. Although it is possible to read this book and prepare and file a patent application, it may still be prudent for you to use the services of a patent attorney to at least help out in the process. But if the cost of this book is the maximum you are willing to pay for legal advice, then keep reading; we will get you through this process yet.

In this chapter, we show you how to construct a patent application that is ready for filing. We describe the various components that make up a patent application and the legal requirements that these components must meet. In addition, we describe what additional documents must be filed along with the patent application and how and where to file the patent application.

Legal-Ease

On June 8, 1995, the United States introduced the concept of a **provisional application** for a patent. Prior to this date, to obtain a filing date you had to file a utility patent application. Today, the normal utility patent application is referred to as either a **conventional patent application** or a **nonprovisional patent application**.

The Parts of the Application

A conventional patent application consists of three main parts: (1) the specification, (2) the claims, and (3) the figures and drawings. Each of these parts must conform to particular requirements regarding formatting and content.

The Specification

The specification is a written technical description of the invention, including how to make or use the invention. The USPTO has proposed a preferred outline to follow for the various sections of the specification. The USPTO is somewhat flexible in the format of a patent application, but, in general, you should follow this outline as close as possible:

- Title of Invention
- Cross-Reference to Related Applications
- Statement Regarding Federally Sponsored Research or Development
- Reference to Sequence Listing, a Table, or a Computer Program Listing Compact Disk Appendix
- Background of the Invention
- Brief Summary of the Invention
- Brief Description of the Several Views of the Drawing
- Detailed Description of the Invention
- Claims
- Abstract of the Disclosure

The USPTO prefers for each of the preceding headings to appear in that order within the patent application with each heading using only capital, nonbolded, and not underlined letters. If a particular heading is not applicable, the USPTO prefers to have the heading listed along with the phrase "Not Applicable" under the heading.

For each of these sections, we tell you what the USPTO requires for the section. But we are also going to give you some tips, hints, and instructions on what you should do in writing each particular section. By the time you finish this chapter, you should be able to prepare a decent patent application that is ready to be filed with the USPTO.

Throughout this chapter, we suggest that you look at patents that are similar to the technology and invention for which you are writing your patent application. To look at similar patents, you can easily access the USPTO website at www.uspto.gov and select Search under the Patents category. If you then select Quick Search, you will be presented with a screen that enables you to enter a term and select a field in which to search. Most of the sections in this chapter correspond directly with a field you can select for the search.

Reviewing other similar patents is a very useful tool when preparing a patent application. Even the most seasoned patent attorney will often do the same thing. And the good news is that patents are public documents (unless they state specifically on their face that the material is copyright protected).

Title of Invention

The title of the invention should appear as the heading on the first page of the specification. Although a title may have up to 500 characters, the title should be as short and specific as possible. When selecting a title, pick something that is short and descriptive. The reason you want to keep it short is that for the next several years, you will probably have to type or write the title many times.

Cross-Reference to Related Applications

In this section, you should identify any other U.S. or foreign patents or patent applications that are related to your patent application. The relationships include claims to priority, continuations, continuations-in-part, divisionals, or simply incorporations by reference. The text in this section must include a reference to each such patent or patent application, identifying them by application number, patent number, or international application number; the filing date; and the nature of the relationship.

The following list presents a few examples:

- If your patent application is going to be based on the priority date of a previously filed provisional patent application, this section should include the following statement:

  This application claims the benefit of the filing date of U.S. Provisional Application for Patent 60/555,555 filed on March 9, 2004.

- If your patent application is a continuation-in-part of a previously filed patent application, this section should include the following statement:

  This application is a continuation-in-part of U.S. Patent Application serial number 09/033,650 filed on March 17, 1993.

- If your application is related to another application, for instance you may be filing two applications for the same product but that includes two different inventions, this section should include the following statement:

  This application is related to U.S. Application serial number __/___,___ filed on the same date as this application.

Legal-Ease

The terms **claim the priority date of** and **claim the benefit of** both refer to identifying the legally recognized filing date for a patent application. The **priority date** is the earliest date that you can claim as the filing date of a patent application. In certain circumstances, you can claim the filing date of a prior application as the filing date for your application. This is a claim to the priority date.

Statement Regarding Federally Sponsored Research or Development

Under certain laws, if an invention is created during a federally sponsored project, the federal government can obtain rights in the invention. In this section, you should include a statement to indicate any rights to inventions made under federally sponsored research and development. A few examples for such statements include the following:

> *This application was supported by NIH Grant No. GM123456. The United States may have rights under this application.*
>
> *This application is part of a government project, Contract No. NAS9-1234567.*

If this section does apply, there are no particular magic words or phrases that have to be used. In general, you simply need to identify that certain aspects of the government may have rights in this patent and the basis for those right.

Reference to Sequence Listing, a Table, or a Computer Program Listing Compact Disk Appendix

In some circumstances, additional material that is relevant or pertinent to the patent application can be submitted on a compact disc. The only disclosure material accepted on compact disc are computer program listings, gene sequence listings, and tables of information. Generally, a compact disc containing such information is submitted when that information is lengthy.

When material is separately submitted on a compact disc, this section of the specification must include a reference to the compact disc and a description of its contents. If a disc contains multiple files or if you submit multiple discs, this section must provide an adequate listing and description of each item. An example of a statement for this section includes:

> *This application incorporates the software program listing included on the compact disk labeled "Additional Materials Disk 1 of 1" including the following source code files:*
>
> *Main Program*
>
> *Compilation Make File*
>
> *Readme File*

When do you need to submit a compact disc? In general, when a printed document becomes burdensome, you should consider submitting a compact disc containing some of the information. However, if you are submitting a computer program that includes more than 300 72-character lines, then you must submit the computer program on compact disc. You can optionally submit a compact disc for computer programs less than 300 lines, for tables of data that would occupy more than 50 pages, or for a gene sequence listing.

Background of the Invention

Regardless of how you structure the application, this section needs to include a statement that identifies the general field in which the invention applies. This statement should not be a mere recitation of the title, but it also does not have to be an elaborate description. For example, one technique to create this statement is simply to paraphrase the patent classifications that are applicable to this invention. Another technique is to simply provide a description of the subject matter of the invention.

When writing this portion of the background section, the best solution is to take a simple two-tiered approach. In the first tier, state a broad category that the invention relates to, and in the second tier, narrow the category. For example, suppose you invented the fridge pack for holding soda cans. You could construct your statement as follows:

> *This invention relates to the field of packaging and, more specifically, to the field of packaging, storing, and dispensing soda cans.*

This section can also include a description of specific problems in the prior art or the current state of the relevant technology that your invention overcomes.

This is the first section of the patent application where you can actually make statements that may result in limiting the scope and strength of your patent. There are two philosophies regarding how this portion of the background section should be constructed. One school of thought is that less is best. The other school of thought takes the position that full disclosure is the safest route.

The less-is-best school of thought operates under the realization that any comparison of your invention to related patents or publications may have the effect of limiting the scope of the claims of your patent. Thus, rather than make such statements at the onset of filing the patent application, this school of thought simply provides a disclosure of the documents at the time of filing or subsequent to the time of filling. If the USPTO then raises any issues regarding the disclosed documents (that is, relies on one of the disclosed documents in a rejection of your claims), then you can address the specifics of that document by making statements regarding the differences between the document and your invention.

The full-disclosure school of thought operates under the realization that a patent is strongest if all of the relevant art is identified and addressed in the patent application.

If you are filing a patent application on your own, you should be very careful about how you compare your invention to related documents. The best and safest approach is to simply list in a narrative fashion the problems that exist in the state of the technology for which your invention will provide a solution. You should also use this section as an opportunity to educate the USPTO regarding the specifics of the technology.

Brief Summary of the Invention

The purpose of the summary section is to present the substance or general idea of the claimed invention in summarized form. The summary should, thus, include a high-level description of the invention that is claimed, and may also include the following information:

- An identification of the advantages of the invention
- How the invention solves previously existing problems, preferably those problems identified in the Background Section
- A statement of the object or purpose of the invention

You should understand that the patent examiners at the USPTO are very busy and under much pressure to examine applications as quickly as possible. Thus, it is likely that a patent examiner may not read your entire patent application. If this is the case, the examiner will most likely focus on the sections that provide him or her with the most information in the shortest period of time. When you write your brief summary, you should write it under the assumption that other than the claims, this is the only section that the patent examiner is going to read.

You should spend a significant amount of time in drafting, reviewing, and perfecting the summary. When you finish writing the summary, read it again and ask yourself the following questions:

- Does the summary disclose the entire invention, including all of the various aspects, in a clear and concise manner?

- Does the summary educate someone with minimal knowledge in the field?
- Are the advantages of the invention clearly articulated?
- Are the problems solved by the invention clearly articulated?

If you can't answer yes to all of these questions, you should go back and revise the brief summary. Also, remember that this is supposed to be *brief* and a *summary*. The emphasis is on keeping it very short and very informative.

Brief Description of the Several Views of the Drawing

If your patent application includes one or more drawings, this section must provide a listing of each figure in the application. The listing should identify each figure by number (for example, Figure 1, Figure 2A, Figure 2B, Figure 3, and so on). In addition, the listing should include a statement that explains what each listed figure illustrates.

There are not too many rigid requirements for this section. You should ensure that the identifying references for the figures are the same throughout the application and the figures. For example, if your drawings use the following format—Fig. 1, Fig. 2, Fig. 3A, Fig. 3B—you should use that same format in this section and throughout the rest of the patent application. In contrast, if your drawings use the following format—Figure 1, Figure 2, Figure 3a, Figure 3b—you should use that format in this section and throughout the patent application.

Detailed Description of the Invention

This section is really the meat of the specification. Within this section you must describe the details of the operation of your invention, including an adequate description on how to make and use the invention. Two main requirements are imposed on this section of the patent application. The overall application must meet these requirements, but generally these requirements are addressed in this section:

- The description has to be enabling.
- The description has to disclose the best mode.

To be enabling, the USPTO requires that the description must provide sufficient detail so that any person of ordinary skill in the pertinent art, science, or area could make and use the invention without extensive experimentation. You may be asking yourself, "How do I know if my description is enabling?" One way to find out is to allow someone else who is skilled in the technical field related to your invention to read your description and then indicate whether he or she believes they could implement the invention based on the description. It is important to ensure that your description is enabling, because if it is not, the USPTO may not allow the patent to be issued, or, if the patent is issued, it may be invalidated later.

What doesn't meet the requirements of an enabling description? First of all, any description of a perpetual motion machine will not be enabling because such a device will never exist. You need to be careful about applying technology in your invention with which you are not very knowledgeable. For instance, suppose you invented an automobile engine that runs on Kool-Aid. It would not be a sufficient description if you simply state that you need to modify the engine so that it burns Kool-Aid rather than gasoline. Instead, you would have to describe the detailed structure of how the engine converts the Kool-Aid into propulsion energy for moving the car.

The best mode for carrying out the invention must be included in the description. For instance, you cannot disclose just one method to implement the invention but keep the best method secret to yourself. In exchange for the monopoly rights that the U.S. government grants to you when it issues a patent, you have to provide a description of the invention that will enable others to benefit from the advancement you have made in the technology. Thus, you must disclose the best mode that you are aware of for implementing your invention. There is no hard and fast rule for determining the best mode. For example, it is not necessarily the cheapest, fastest, strongest, most commercially feasible mode. But these factors could be weighed in the mind of the inventors for determining the best mode.

The detailed description must include a description of each drawing, and each element in the drawings should be mentioned in the description.

In formatting the detailed description, we suggest using the following technique: In the opening paragraph, restate the essence of the invention in one or two sentences and then list the various aspects or advantages of the invention.

In the opening of the second paragraph, enter the text: "Now turning to the figures in which like references refer to like elements throughout the several views, various aspects and embodiments of the present invention are described."

From this point, you should describe the details of each figure and provide further details regarding the invention. You should make sure that each element in the figures that includes a reference number is described in the detailed specification. In addition, you should strive to write the text in such a manner that someone could re-create the drawing based on the description of the drawing. In some instances this may simply not be possible, but in most cases it should be quite doable.

Abstract of the Disclosure

In the patent application, the abstract appears near the end of the application, right before the figures. However, in a published patent, the abstract appears on the front page. The abstract is a concise description of the invention that allows the public to quickly determine the nature of your invention. The length of the abstract is limited to 150 words.

The Claims

The claims of a patent application identify the exact boundaries of what is protected by the patent. This is absolutely the most complicated step in the process of preparing a patent application. A claim is a single sentence that captures the entire essence of at least one aspect of your invention. It is very important to carefully draft your claims because a single word, or even a letter or punctuation mark, can greatly limit the scope of your claims.

There are two types of claims: independent claims and dependent claims. An independent claim stands by itself. A dependent claim is attached to an independent, includes each of the elements of the independent claim, and then adds one or more additional elements or limitations.

The patent application must include at least one claim, but it can have as many claims as you desire. However, it should be noted that the filing fee for a patent application only includes the cost for 20 claims, with 3 of those claims being independent. Any claims beyond 20 and any independent claims beyond 3 will result in an additional charge.

Independent Claims

An independent claim includes a preamble followed by a listing of elements critical to the invention. There are several types of independent claims, but generally they can be categorized in one of the following three classes:

- Apparatus claim
- System claim
- Method claim

The following list provides an example of a preamble for a claim in each of these classes:

> *An apparatus for (short description of what is accomplished), the apparatus comprising:*
>
> *A system for (short description of what is accomplished), the system comprising:*
>
> *A method for (short description of what is accomplished), the method comprising the steps of:*

Following the preamble is a list of elements defining the invention. In an apparatus claim, these elements are parts or features of the apparatus. For example, an apparatus claim for a toothbrush could include the following parts:

- A handle
- A rubber sleeve to go over one end of the handle to provide a grip
- A plurality of soft bristles grouped together on one end of the handle

In addition, the apparatus claim for the toothbrush could include the following features:

- A rounded end to prevent injury to the mouth
- A curve in the handle to facilitate reaching the backside of the teeth

In a system claim, the elements are parts or components of the system, or features of those components. In a method claim, the elements are steps involved in the procedure or process. For instance, in a method claim for a method for brushing your teeth, the claim could include the following steps:

- Wetting the brush
- Applying toothpaste to the brush
- Opening the mouth
- Inserting the brush with toothpaste into the mouth
- Applying the brush to one or more teeth and moving the brush back and forth across the teeth

Dependent Claims

Dependent claims are used to further limit the scope of an independent claim. Thus, a dependent claim will include or incorporate each of the elements of another claim and add one or more new elements or augment one or more elements. As an example, a dependent claim for the previously listed apparatus claim for the toothbrush could add a further limitation as shown here:

> *The apparatus of claim (insert the number of the claim that this claim is related to), wherein the rubber sleeve includes a plurality of grooves.*

If your goal is to obtain a patent with the broadest claims possible, your application should include broad independent claims that are then narrowed by one or more dependent claims. If the USPTO rejects one of your independent claims because they have found a reference that discloses each element or otherwise, they may allow one or more of your dependent claims if they include elements that were not found in any references uncovered by the USPTO.

Multiple Dependent Claims

Some patent applications will include a dependent claim that can depend from one or more other claims. In such a claim, the dependency must refer to the other claims in the alternate form only. For instance, if you have a claim that can be used to limit two other claims (for example, claims 1 and 5), the multiple dependent claim should state:

> *An apparatus as in claims 1 or 5, further comprising …*

Legal-Ease

The terms **comprising** and **includes** are very different in terms of their use in a patent claim. The term *includes* is interpreted as a limiting list of elements. The term *comprising* is used to broaden a claim. Comprising means something includes but is not limited to the listed elements.

Drafting the Claims

It is good practice to draft your claims in a way that describes your product, techniques to design your product, and if possible, your competitor's products. Ask yourself the following questions:

- Do my claims fully describe my product?
- Would it be easy to modify or eliminate one of the elements of my claims?
- Do the claims cover potential advances in technology that are anticipated or likely?
- Do the claims cover the products of my competitors?

When you are drafting your claims, one technique you should consider is creating a matrix or a claim chart. Going down the left side of the matrix, you should list each of the various elements and aspects of your invention. Across the top edge of your matrix, you should sequentially number the columns. These numbers will represent the claims in the application.

Then you should group these elements and aspects in accordance with their importance to the core of your invention (that is, critical elements). These groupings should help identify the elements to be included in the independent claims. You should strive to ensure that your independent claims only include the critical elements and include as few elements as possible necessary to separate your claim from the prior art information already available to the public. Alternatively, you could have one independent claim that includes the minimum number of critical elements and two additional independent claims that each include the minimum number of critical elements and at least one additional critical or noncritical element.

You should try to limit the number of claims in your application to the 20 claims (with 3 independent claims) that are provided for in the basic filing fee. If necessary to fully claim your invention, use more claims, but always try to limit the claims to as few as possible. The reason for this is that at the Patent Office, an examiner gets paid the same for reviewing a patent with 20 claims as he does for reviewing a patent with 200 claims. Obviously, the examiner is going to be much more willing and eager to review the patent that only includes 20 claims over the patent that has 200 claims. Our rule of thumb is that if you can't fully claim an invention in around 20 claims, then you probably don't fully understand the invention.

The patent examination process is a very lengthy process that can be shortened or extended greatly based on the examiner and his interest and desire to review your patent application. Thus, you want to do everything you can to make your patent application attractive to the examiner.

The Figures or Drawings

If necessary to describe and understand the invention, a patent application must include drawings.

As a rule of thumb, you should ensure that the drawings illustrate each and every feature, aspect, or element of the invention as specified in the claims. By failing to include a drawing, the USPTO may consider the application to be incomplete.

The patent application can include as many drawings as necessary to fully describe the invention. The drawings can include cross-sectional views, exploded views, perspective views, and partial views.

When to Involve a Patent Attorney or Agent

Somewhere along the process you may decide to throw up your hands and hire a patent attorney. There are advantages and disadvantages regarding what stage you should solicit the help of a professional. In this section, we provide some guidance regarding this issue.

The Attorney's Liability

You should understand that as soon as an attorney gets involved, the attorney is subjected to liability regarding any further actions. Because of this liability, the typical attorney will not want to be in a position to co-write the patent application with you.

For example, suppose you wrote the entire patent application and handed it over to the attorney for filing but told the attorney not to change the application whatsoever. Suppose that down the road, it is determined that the specification is faulty and does not adequately disclose the invention. The attorney has opened himself or herself to tremendous liability at that point. If the attorney were to accept such a task, the attorney should document the facts very carefully to ensure that it is very clear you have requested him or her to file the application "as is" without any review. From the attorney's perspective, you should either bring the attorney onboard to draft the entire patent application or wait until you have filed the patent application. At this point, the attorney can take over the prosecution of the patent application.

Don't Jump the Gun

Don't make the mistake of bringing the patent attorney onboard too early! Once that attorney's clock starts ticking, it can be difficult to stop it. Do your homework first. Make sure your invention is patentable. You can easily get online and search for items that may predate your date of invention. You can also visit the USPTO website (www.uspto.gov) and search for other patents that may indicate that someone else has already invented your invention. You can even hire a professional searching firm.

If you feel solid about going forward with your invention after having done all this homework, are you ready to bring an attorney onboard yet? No, not yet. The next thing you should do is sit down and write out a thorough description of your invention. The description does not have to conform to any particular format, but you should include the following information:

- A clear description of what the invention is
- How to make the invention or use the invention
- Why the invention is advantageous over other available options
- What other solutions are currently available and why they are not as good as your invention

You should also sketch drawings that show how your invention is constructed or operates.

This is the earliest time you should consider bringing a patent attorney onboard. At this point, you are ready to hand over a complete package of information that the attorney can convert into a patent application.

A Good Transition

There are a few points in the process of preparing, filing, and prosecuting a patent that are good transition points. There are advantages and disadvantages for transitioning at each point. The following list summarizes these transition points:

- **In the beginning.** Lay the foundation of a good, solid description of the invention before you bring in an attorney. This will help you save money.

- **After the filing.** Once you file the application, you can hand over the prosecution to a patent attorney and the attorney will not be unfairly faced with a huge liability. At this point, the attorney may begin charging you periodically for maintaining the docket (or watching the progress) for this application. This is a minor cost, and it can give you great peace of mind to know that now it is someone else's responsibility to make sure that the necessary steps are taken.
- **Upon receiving an office action.** It can be as long as two and a half years after filing and application before you get an office action. Thus, you may want to wait until you receive your first office action before bringing an attorney onboard.
- **Before the patent issue.** If you actually prepare, file, and prosecute the patent application to allowance all by yourself, kudos to you. However, before you pay the issue fee for the patent, you may want to consider having a patent attorney review everything that has transpired. It is quite possible that you may have obtained a patent that can be easily designed around or your claims may be so narrow that the patent is worthless. Prior to paying the issue fee, you can possibly fix these deficiencies. Fixes to the problems could include filing a continued prosecution or a divisional patent application.

The Least You Need to Know

- A patent application is a complex technical document, and if you do not draft it properly it can limit your rights.
- The claims of a patent define your rights in the invention.
- You can write, file, and prosecute a patent application on your own.
- If you are going to use a patent attorney, you should bring the patent attorney onboard at the appropriate time.

Chapter 4

The Details of the Conventional Patent

In This Chapter

- The detailed formatting requirements for the specification
- Formatting and numbering the claims
- Preparing the figures
- Assembling the entire package for mailing
- Filing the patent application

If you thought writing the patent application was the hard part, wait until you learn how much work you have to do in formatting the application and preparing the filing documents. The USPTO has specific formatting requirements for patent applications—and justifiably so. The formatting requirements are designed to help the Patent Office more efficiently process the patent applications.

In this chapter, you learn the major formatting requirements for preparing a patent application. In addition, you learn what forms need to be filed with a patent application and how to fill out those forms. Finally, you learn how to assemble a final package for filing and how to file the patent application with the Patent Office.

Formatting the Patent Application

This section outlines the formatting requirements for a patent application.

The Format of the Specification

The specification, which includes everything but the claims and the figures and drawings, must meet each of the formatting requirements:

- **English language.** The patent application should be in English. If the patent application is in a foreign language, the application must include a verified translation into English.
- **Typewritten.** All of the papers in the patent application must be typewritten or produced by a mechanical (or computer) printer and in black ink.
- **Paper.** The papers filed in a patent application should only include text on one side; should have a portrait orientation; and the paper should be white, flexible, strong, smooth, nonshiny, durable, and free from

any holes, creases, tears, or cracks (with the exception of the holes that the staple creates).

- **Paper size.** Each of the papers filed in the patent application should be the same size. Two sizes are acceptable to the Patent Office: (1) 8.5 by 11 inches (21.6 cm by 27.9 cm) or (2) DIN size A4 (21.0 cm by 29.7 cm).
- **Margins.** The left edge should include a 1-inch (2.5 cm) margin and the top, right, and bottom edges should include a ¾-inch (2.0 cm) margin.
- **Font size.** The text of the patent application should be at least 12-point size.
- **Line spacing.** You should use one and a half or double line spacing throughout the document.
- **Headers.** The headers should be in uppercase letters, left-justified, and not bolded or underlined.

The Format of the Claims

Each of the claims should be numbered sequentially using Arabic numbers. Each dependent claim must refer to at least one other claim using the Arabic number associated with that claim. Each element of the claim should be indented to separate that element from the preamble and other elements.

The claims should be grouped in accordance with their relationship to each other. In addition, you should order the claims so that each claim is adjacent to the claim from which it depends or other claims that depend from the same claim. Claims should only depend from claims that have a lower claim number.

Let's look at some examples. Suppose you have two independent claims (claims 1 and 6) and eight dependent claims (claims 2 to 5 and 7 to 10). Here is the preferred ordering for the claims, along with the identification of their dependencies:

Claim 1 (independent claim)

Claim 2 (depends from claim 1)

Claim 3 (depends from claim 2)

Claim 4 (depends from claim 1)

Claim 5 (depends from claim 4)

Claim 6 (independent claim)

Claim 7 (depends from claim 6)

Claim 8 (depends from claim 7)

Claim 9 (depends from claim 8)

Claim 10 (depends from claim 6)

Here is an example of an ordering format that is not preferred:

Claim 1 (independent claim)

Claim 2 (depends from claim 3)

Claim 3 (depends from claim 1)

Claim 4 (independent claim)
Claim 5 (depends from claim 2)
Claim 6 (depends from claim 4)
Claim 7 (depends from claim 10)
Claim 8 (depends from claim 1)
Claim 9 (depends from claim 4)
Claim 10 (depends from claim 9)

The Format of the Figures and Drawings

If you decide to prepare the patent application by yourself, you may want to consider hiring a professional patent draftsman for preparing your drawings. The requirements imposed on the drawings are quite extensive, and until you satisfy each of the requirements, the USPTO will not issue your patent.

The Paper Requirements

The papers filed should conform to the same page size and characteristics used in the specification (see "The Format of the Specification" earlier in this chapter) with the following restrictions:

- **Margins.** The margins for each page must be structured as follows:

Top margin	1 inch	2.5 cm
Left side margin	1 inch	2.5 cm
Right side margin	⅝ inch	1.5 cm
Bottom margin	⅜ inch	1 cm

- **Frames.** The sheets should not contain frames around the usable surface, but should have scan target points (crosshairs) printed on two opposite margin corners.

The Drawings

Generally, drawings are required to be submitted in black ink on white paper. The ink should be India ink or its equivalent. Computer-generated printouts should be originals rather than photocopies.

If you insist on submitting a color drawing rather than a black-and-white drawing, it will cost you. Basically, you will be required to submit a petition to the USPTO including the following:

- An extra fee (we told you it would cost you)
- Three sets of the color drawings
- Following the heading "BRIEF DESCRIPTION OF SEVERAL VIEWS OF THE DRAWINGS" you must include the statement:

  The patent or application file contains at least one drawing executed in color. Copies of this patent or patent application will be provided by the Office upon request and payment of the necessary fee.

The USPTO generally only permits photographs to be submitted when the invention is not capable of being illustrated in an ink drawing or where the invention is shown more clearly in a photograph. When permitted to file photographs, only one set of black-and-white photographs is required, and no additional fee is required. Photographs must be presented in accordance with the same sheet size requirements as other drawings. The photographs must be of sufficient quality so that all details in the drawing are reproducible in the printed patent or any patent application publication. Color photographs will be accepted in patent applications if the conditions for accepting color drawings have been satisfied.

Identification of the Drawings

Each drawing page should include information sufficient to identify the drawing, should it become separated from the rest of the patent application. Thus, you should place the following information on the face of each drawing:

- Title of the invention
- Name of the inventor(s)
- The serial number of the application (if known)
- Docketing number
- Name and phone number to call

This information should be centered inside the top margin of each drawing, outside of the drawing area.

Lettering, Numbers, and References

The drawings should include reference numbers, sheet numberings, and descriptive text when necessary. For all such letters and numbers, the font size of the characters must measure at least ⅛ inch (0.32 cm) in height. The characters should not be placed in a manner that they cross or overlay on top of any lines or hatched or shared surfaces.

The English alphabet should be used for all characters unless the character is used to depict a special meaning such as angles and variables in mathematical formulas. The orientation of all characters should be the same as the overall layout of the drawing.

Every reference character included in a drawing must be mentioned in the specification.

Lead Lines

Labels used as reference characters are associated with certain aspects of the drawing through the use of lead lines. The lead lines may be straight, curved, or bent and should be as short as possible. The lead lines must extend from the proximity of the reference label to the feature associated with the reference label. All reference labels must include a lead line if at all practical. If it is not practical to use a lead for a reference label, the reference label should be placed close to or within the element it is associated with, and it should be underlined.

The Filing Documents

The patent application must be filed along with other important documents. Some of the documents are so important that if they are missing, you will not be granted a filing date and you will have to submit the missing documents to obtain the filing date.

What to File

When filing a patent application, it is important to ensure that you meet the minimum requirements for obtaining a filing date. If you do not meet the minimum requirements, you may lose valuable rights in your invention or someone else may obtain rights in your invention during the interim period.

The minimum requirements for filing a patent application include a completed specification that is enabling and discloses the best mode, at least one claim, and any required figures.

To keep the filing date, additional information and fees must be filed with the Patent Office in a timely manner. Ideally, when you file a patent application, you should include the following items along with the patent application:

- Patent transmittal sheet
- Fee transmittal sheet
- Fee payment
- Oath or declaration

- Assignment recordation cover sheet and supporting documents (if any)
- Appointment of a power of attorney (if necessary)
- Application data sheet
- Return receipt postcard

Patent Transmittal Sheet

The transmittal can be the standard Utility Patent Application Transmittal Form available from the USPTO website (Form PTO/SB/05) or a custom transmittal letter. In either case, the transmittal should provide the following information:

- The name of the applicant
- The title of the invention
- The Express Mail tracking number
- A listing of the items being filed with the application
- If the application is a continuation, the type of continuation along with the related patent application number and the examiner assigned to that application
- The correspondence address for communicating with the applicant or the applicant's attorney
- The signature and date of the applicant(s) or the authorized person filing on behalf of the applicant(s)

Fee Transmittal Sheet

Whenever you pay any fees to the USPTO, the fee should be attached to a paper that identifies what the fee should be applied to, as well as other important information. The USPTO provides a standard form called the Fee Transmittal Sheet that can be downloaded from the USPTO website at www.uspto.gov (Form PTO/SB/17).

Good Counsel

If a patent application has more than one inventor, you need to select an order for listing the inventors. The application is generally referred to by the last name of the first inventor. For instance, if inventors are listed in the order of Dolores Smith, Fran Jones, and Melisa Baise, then the application and the issued patent would be referred to as Smith *et al.*

In place of the Fee Transmittal Sheet, you can prepare your own letter. In either case, your payment should be accompanied by a paper that provides the following information:

- The filing date
- The name of the first listed inventor
- A docketing number associated with the application
- Whether the applicant is a large or small entity

- What fees are being paid (filing fee, excess claims)
- The total amount of the fee payment
- The method of payment
- An authorized signature

Fee Payment

When you file your patent application, you should submit the filing fee with the patent application. The filing fee can be paid using one of three methods:

- A check payable to the Commissioner for Patents and Trademarks
- A deposit account (usually only used by attorneys)
- Credit card payment

If you decide to pay the fees using a credit card, it is important to use the Credit Card Payment Form provided by the USPTO and to list your credit card information only on this form. This form does not become a part of the file history of the patent, whereas any other document does. Thus, if you put your credit card authorization and information on any other document, your credit card information can be obtained by anyone viewing the file history of your patent application. The Credit Card Payment Form can be downloaded from the USPTO website (Form PTO 2038).

What is this large entity and small entity thing all about? The USPTO provides a benefit for individual inventors or small companies (companies that have less than 500 employees). The USPTO classifies individual inventors and small companies as small entities. As a small entity, most of the fees assessed for filing and prosecuting a patent application are reduced by 50 percent.

Declaration

The patent application should include a declaration, signed by each inventor, that states …

- That the residence, mailing address, and citizenship information provided for each inventor is correct.
- That the signing parties believe the listed inventor(s) to be the original and first inventor(s) of the subject matter that is claimed and for which a patent is sought.
- That the signing parties have reviewed and understand the contents of the above identified specification, including the claims, as amended by any amendment specifically referred to above.
- That the signing parties acknowledge the duty to disclose information that is material to patentability of the invention.

The declaration must also identify with as much specificity as possible the patent application, a correspondence address, and any foreign patent applications.

The USPTO provides a standard form called the Declaration for Utility or Design Patent Application that you can download from the USPTO website (Forms PTO/SB/01, PTO/SB/2A, and PTO/SB/2B).

Assignment Recordation Cover Sheet and Supporting Documents

In some instances, you may want to assign the rights in your patent application to another entity or individual. This information can be conveyed to the USPTO at the time of filing the patent application; however, the assignment can also be filed at a later date.

If you are going to file an assignment, you need to file a copy (preferably the original) assignment document, an assignment recordation cover sheet, and a recordation fee. If you file the assignment together with the patent application, you can use the same Fee Transmittal Form; however, if you file the assignment at a later date, you will need to include a Fee Transmittal Form.

The assignment recordation cover sheet identifies the nature of the documents to be recorded and the entities involved (for example, the person assigning the patent rights and the person or entity receiving the patent rights).

The USPTO provides a standard form called the Recordation Form Cover Sheet that can be downloaded from the USPTO website (Form PTO 1595).

Once the assignment is recorded, the USPTO will return your documents to you.

Appointment of a Power of Attorney

You can get an attorney involved in the process at any stage. If and when you bring an attorney on-board, you will need to file an appointment of a power of attorney.

The USPTO provides a standard form for appointing a power of attorney that can be downloaded from the USPTO website (Forms PTO/SB/80 and PTO/SB/81).

Return Receipt Postcard

When you file anything with the USPTO through the mail, you should include a return receipt postcard. One side of the return receipt postcard should include a description of each item that is included in the filing package, the number of pages for each item, and information to identify the patent application. The other side of the postcard should be self-addressed and include sufficient postage.

Upon receipt at the USPTO, the list on the postcard will be compared to the actual contents of the delivery. Any discrepancies between the detailed list and the actual contents will be noted on the poscard. The postcard will be initialed and date-stamped by the person at the USPTO who received the delivery and will be returned by mail to the addressee whose name appears on the postcard.

Application Data Sheet

The patent application may include an Application Data Sheet. Instructions for completing the Application Data Sheet can be obtained from the USPTO website at www.uspto.gov/web/offices/pac/dapp/sir/doc/patappde.html.

The information provided in the Application Data Sheet is presently being provided in multiple parts of patent applications—specifically the ones listed earlier in this chapter. The Application Data Sheet serves to group all of this information onto a single form. Ultimately, the Application Data Sheet assists the USPTO in capturing the data about the patent application. If you use the Application Data Sheet, some of the previously mentioned forms can be substituted with simpler forms. These simpler forms are also available on the USPTO website.

Assembling the Patent Application

Prior to mailing the patent application to the USPTO, you should assemble the package and staple all of the sheets together.

The following format is a suggested technique for assembling the package (from top to bottom). However, before stapling everything together, you should make sure that all of the papers are signed and you should make a copy of everything for your own records.

- Patent Transmittal Sheet; the postcard should be stapled to the top and front of the Patent Transmittal Sheet
- Fee Transmittal Sheet; the check or credit card payment form should be stapled to the Fee Transmittal Sheet
- Application Data Sheet (if used)
- Oath or Declaration
- Appointment of a power of attorney (if necessary)
- Assignment recordation cover sheet and supporting documents (if any) all stapled together
- The specification and the claims
- The figures and drawings stapled together

Finally, a single staple should be used to staple the entire package together. A binder clip may also be used, but a staple is preferred.

Where and How to File

You should file your patent application using Express Mail. The reason for this is that for any item filed using Express Mail, the USPTO will recognize the date that the package was stamped by the U.S. Post Office as the date of receipt by the USPTO.

The package should be sent to the following address:

> Mail Stop Patent Application
> Commissioner for Patents
> PO Box 1450
> Alexandria, VA 22313-1450

The Least You Need to Know

- The patent application must conform to detailed formatting requirements.
- Each claim should be sequentially numbered, and dependent claims should be grouped with the independent claim to which they relate.
- Patent applications should be filed via Express Mail.
- The patent application must be filed along with other forms required by the USPTO, or at least equivalents to those forms.

Chapter 5

The Design and Plant Patent Applications

In This Chapter

- The parts of a design patent application
- The parts of a plant patent application
- How to file a design or plant patent
- Remedies available for infringement

In general, when people think about patents, they think about particular types of inventions. For example, if we were sitting in your family room and asked you to look around the room and point out things that could be patented, what types of things would you select? You would probably point to the television, the stereo, and the remote control. When we think of patents, we think of machines, electronics, computers, telephones, and so on. However, there are two things, or types of things, that can be patented that we bet you did not know about: designs and plants.

Take a moment to look at the outfit you're wearing (unless you are a guy—we don't really wear outfits, just clothes). Do you like the way your clothes look? Would you ever think that you could patent the way a pair of pants look? Or if you have a cellular telephone in your pocket or purse, pull it out. Do you like the way it is shaped and the way the keys are laid out? Would you ever believe that you could patent the way a cellular telephone looks? Well guess what, you can! You will learn in this chapter that you can actually patent the ornamental aspects of an article. This is a design patent.

Now, let's think about high school days. Maybe this didn't happen to you, but it may have happened to someone near you, or maybe it just happened to me (we won't say which of us is me). But nonetheless, it is a good example. At the end of the school year, along with turning in the textbooks, retrieving confiscated yo-yos and hats from the principal's office, and cleaning out your desk, we always had the infamous cleaning out of the gym locker. So as you pull out the empty cans of deodorant and near-empty bottles of shampoo and conditioner, you reach to the far back and you finally find that source of that particular odor that had been haunting you since shortly after you had lost a pair of your gym shoes—yes, it was your long-lost gym shoes. As you pull them out of the back of the locker and peek inside, you find a green and yellow fungus has decided to take root and establish a home right there in your long-lost black Converses. Well guess what—if this is a newly discovered plant and you can reproduce it without using seeds, then you could actually obtain a patent for this plant. This is referred to as a plant patent.

In this chapter, we explain what design and plant patents are, how to file and obtain such a patent, and what types of protection and rights that owning such a patent gives you. We promise that we will not have any more gross gym shoes examples.

The Design Patent

Let's look at the definition of a design patent. Avoiding all of the legal mumbo-jumbo, a design patent is simply a way for someone to prevent others from copying the visual characteristics of an article. What are the visual characteristics? They would include the shape, the ornamentation applied to the surface of an item, or both. In general, you are looking at protecting the visual impression that is created by looking at an item. Thus, a design patent enables you to protect the way your product or item of manufacture looks. This is very different from a utility patent, which protects the way that your product operates. Utility patents focus on protecting function; design patents focus on protecting ornamentation.

Examples of Design Patents

The following list provides some examples of designs that can be protected through a design patent:

- United States Design Patent Number D487,579 protects the ornamental design of a clipboard.
- United States Design Patent Number D390,312 protects the ornamental design of a dog-shape sprinkler.

From these examples, you can see that there is quite a variety of designs that can be protected through the use of a design patent.

Just the Facts

The Coca-Cola Company has more than 55 design patents that protect the ornamental aspects of its bottles or items related to its bottles.

The Rights of a Design Patent Owner

If you obtain a design patent, what rights do you get? You get the right to prevent others from making, using, selling, or importing an infringing product within the United States. This right is granted for a period of 14 years from the date that the design patent is granted.

What is an infringing product? The courts have said that infringement is based on a substantially similar test. Thus, in the eye of an ordinary observer giving such attention as a typical purchaser would give, would a potentially infringing design be so similar to a patented design that the purchaser could be deceived into purchasing the infringing product, thinking it to be the patented product? If this is the case, then courts have concluded the products are substantially similar and infringement is occurring.

The Design Patent Application

The design patent application can be viewed as including three parts: the specification, the claim, and the drawings.

The specification is divided into several sections:

- **Preamble.** The preamble lists the name of the applicant, the title of the design, and a brief description of the nature and intended use of the item for which the design is intended.
- **Cross-reference to related applications.** This section identifies any applications that are related to the present application, and is similar to the corresponding section in a conventional patent. (More detail is available in Chapter 3.)
- **Statement regarding federally sponsored research or development.** This section is similar to the corresponding section in a conventional patent (see Chapter 3).
- **Description of the figure or figures of the drawing.** This section basically describes the type of view or views presented in the drawing or drawings. For

example, typical views include bottom views, top views, left or right side views, and perspective views.

- **Feature description.** This section describes what is not included as the claimed invention. For example, if the invention is only a portion of an overall device, such as the bill of a cap, then it may be necessary for the drawings to show the entire cap to help gain an overall understanding of the invention. Oftentimes, this is done by using broken lines on the drawing to illustrate portions of the product that are not included as part of the invention.

If the drawing is lined for color, the feature description must use the following statement:

> *The drawing is lined for color.*

If the drawing uses broken lines to indicate an environment in which the claimed design exists, the feature description must use the following statement:

> *The broken line showing of (insert name of structure) is for illustrative purposes only and forms no part of the claimed design.*

A design patent includes only a single claim; more than one claim is not permitted. The claim is a very general statement that usually takes a form similar to the following example:

> *The ornamental design for a (insert name of article), as shown and described.*

A Picture Is Worth a Thousand Words

The bulk of a design patent application is in the drawings, and every design patent application must include either a drawing or a photograph of the claimed design. The drawing or photograph constitutes the entire visual disclosure of the claim. Thus, it is very important that the drawing or photograph be clear and complete, and that nothing regarding the design sought to be patented is left to the imagination.

The design to be protected must be represented in the drawings by a sufficient number of views to constitute a complete disclosure of the appearance of the design, including front, rear, top, bottom, side, and perspective views. Views that are redundant of other views of the design or that are flat and include no ornamentation may be omitted from the drawing if the specification makes this explicitly clear. For example, if the left side and the right side are identical or mirror images, rather than including duplicative drawing, only a single drawing can be submitted and the specification can simply state that the other side is identical or a mirrored image of the illustrated side.

Surfaces of the drawing may be shaded to show character and contour of any three-dimensional aspects of a design or to distinguish between open and solid areas of an item. A contrast in materials may be shown by using line shading and stippling to differentiate between the areas. This technique allows for a broad claim that is not limited to particular colors or materials, but rather just a contrast in materials.

Broken lines can be used in the drawings to disclose the environment related to the claimed design, define the boundaries of the claim, or both. The environment related to the claimed design can be included when it is necessary to describe more clearly the design. For example, consider a golf club–related design patent for the contour on the back of the club head. The entire golf club head can be included as an environment for a contour on the back of a golf club head. Only the contour will be considered a part of the claimed invention. Broken lines can also be used to establish a boundary within which the claimed design exists. However, broken lines cannot be used to indicate that items are of less importance in the design or to show hidden planes or surfaces.

The specific formatting for design patent applications drawings is the same as for utility applications described in Chapters 3 and 4.

The Filing Documents

After you have completed your design patent application, you are ready to assemble the package and mail it to the USPTO. When you submit your application, you should go through the following checklist to make sure that you have everything necessary to secure a filing date. When assembling your package for mailing, you should attach the following items in this order:

- ❑ Design application transmittal form. The transmittal form can be downloaded from the USPTO's website at www.uspto.gov/web/forms/sb0018.pdf.
- ❑ Fee transmittal form.
- ❑ Fee payment.
- ❑ Application data sheet.
- ❑ Specification.
- ❑ Drawings or photographs.
- ❑ Executed oath or declaration.

The fee transmittal form, fee payment, application data sheet, and the executed oath or declaration are further described in Chapter 4. Similar to filing a conventional patent (as described in Chapter 4), a self-addressed and stamped return receipt postcard should be included along with the design patent application to serve as confirmation of delivery.

How to File the Design Patent Application

Design patent applications should be filed using Express Mail available from the USPS. The application should be mailed to the following address:

> Mail Stop Design
> Commissioner for Patents
> PO Box 1450
> Alexandria, VA 22313-1450

Design patent applications cannot be filed electronically.

The Plant Patent

Let's look at the definition of a plant patent. You can file for a patent for any distinct and new variety of plant that you invent or discover and can asexually reproduce. This includes cultivated spores, mutants, hybrids, and newly found seedlings, other than a tuber-propagated plant or a plant found in an uncultivated state. If all the requirements for filing and for patentability are met, then you can obtain a plant patent.

The Rights of a Plant Patent Owner

If you obtain a plant patent, what rights do you get? You get the right to prevent others from asexually reproducing the patented plant, and from using, offering for sale, or selling the asexually reproduced plant, or any of its parts, throughout the United States. In addition, the patent owner can exclude others from importing into the United States the asexually reproduced plant or any of its parts for a period of 20 years from filing the plant patent application.

The Plant Patent Application

The plant patent application can be viewed as including three parts: the specification, the claim, and the drawings and specimen.

The specification is divided into several sections:

- Title of the invention.
- Cross-reference to related applications.

- Statement regarding federally sponsored research or development.
- Latin name of the genus and species of the plant claimed.
- Variety denomination.
- Background of the invention.
- Brief summary of the invention.
- Brief description of the drawing.
- Detailed botanical description.
- Abstract of the disclosure.

These sections are similar to those in a utility application and are described in more detail in Chapter 3.

The specification of the plant patent application must describe in detail the botanical characteristics of the plant that distinguish the plant from related known varieties and its antecedents. This description must be expressed in botanical terms in the general form followed in standard botanical textbooks or publications dealing with the varieties of the kind of plant involved (evergreen tree, dahlia plant, rose plant, apple tree, and so on), rather than a mere broad nonbotanical characterization such as commonly found in nursery or seed catalogs. The specification should also include the origin or parentage and the genus and species designation of the plant variety sought to be patented. The Latin name of the genus and species of the plant claimed should be included under the appropriate heading.

The specification must particularly point out the location or place of business and in what manner the variety of plant has been asexually reproduced.

A plant patent application must include a single claim. The claim formally recites a claim to the new and distinct variety of the specified plant as described and illustrated and may also recite the principal distinguishing characteristics of the plant. Only one claim is allowed in a plant patent. The general structure of a plant patent claim is as follows:

> *A new and distinct variety of (insert plant category), substantially as illustrated and described, that is further characterized by (insert listing of further characteristics that describe the plant such as blooming times, frequency, colors, and so on).*

The drawings in a plant patent application must comply with the same formatting requirements as described in Chapters 3 and 4 for a utility patent application. In addition, the drawings may be submitted in color if the color is a distinguishing characteristic of the plant. The drawings must disclose all of the distinctive characteristics of the plant that can be reasonably represented visually. If the drawings are submitted in color, two color copies must be provided and preferably a black-and-white copy should be included, although it is not required.

In some cases, the USPTO may require specimens of the plant that is the subject of an application. When requested by the USPTO, you must furnish specimens of the plant, or its flower or fruit, in the quantity and at a time in its stage of growth as may be designated for study and inspection. When requested, the USPTO will provide you with instructions for packing and shipping the specimens. If it is not possible to forward a requested specimen, you must make plants available for official inspection at the location where they are grown. For example, if you have just invented and filed an application on a new form of a very tall tree, do not send it in to the USPTO unless requested—surely they would not appreciate that.

The Filing Documents

After you have completed your plant patent application, you are ready to assemble the package and mail it in to the USPTO. When you submit your application, you should go through the following checklist to make sure that you have everything necessary to secure a filing date. When assembling your package for mailing, you should attach the following items in this order:

- ❑ Plant application transmittal form. The transmittal form can be downloaded from the USPTO website at www.uspto.gov/web/forms/sb0019.pdf.
- ❑ Fee transmittal form.
- ❑ Fee payment.
- ❑ Application data sheet.
- ❑ Specification.
- ❑ Drawings or photographs.
- ❑ Executed oath or declaration.

The fee transmittal form, fee payment, application data sheet, and executed oath or declaration are further described in Chapter 4. As is the case with filing a conventional patent as described in Chapter 4, you should include a self-addressed and stamped return receipt postcard along with the plant patent application to serve as confirmation of delivery.

How to File the Plant Patent Application

Plant patent applications should be filed using Express Mail available from the USPS. The application should be mailed to the following address:

Mail Stop Plant
Commissioner for Patents
PO Box 1450
Alexandria, VA 22313-1450

Like design patent applications, plant patent applications cannot be filed electronically.

The Least You Need to Know

- The key element in conveying what is protected for a design patent is the figures.
- A plant patent only provides limited protection; a utility patent can also be obtained for a plant patent to gain further rights.
- Asexual reproduction is protected through a plant patent; a PVPA registration provides protection against sexual reproduction.

Chapter 6

The Provisional Patent Application

In This Chapter

- The benefits of provisionals
- What is and is not included in provisionals
- How to avoid the pitfalls of provisionals
- How to give away your invention

Guess what? The USPTO has a heart! In 1995, they started accepting something called a provisional application for a patent. A provisional application is most often described as a preliminary patent application—a placeholder until you have the ability to file a conventional or nonprovisional application.

The benefits of filing a provisional application are numerous. The requirements of what must be contained are minimal, the cost is significantly less than that of a nonprovisional application, and the time it takes to prepare and file a provisional application can be much shorter. However, you should be aware that there are many potential pitfalls and risks when filing a provisional application.

This chapter will help you understand the background of provisional applications, when it makes sense to file them, and the problems that you might face with your provisional. We'll also provide you with some helpful hints if you want to file a provisional application by yourself. Finally, we'll go into a couple of other preliminary filings that you may want to consider: the Statutory Invention Registrations and Invention Disclosure Documents.

Provisional vs. Nonprovisional Applications

As you may have gathered, the USPTO has two different categories of applications: provisional and nonprovisional (also known as conventional) applications. However, it is important to understand that a provisional application for patent is not really an application whatsoever. The USPTO does not review a provisional application for patentability, and the USPTO cannot issue a patent based solely on a provisional application. In essence, the provisional patent gets stored in a shoebox at the USPTO and simply sits there until one of two events occurs:

- The provisional application is abandoned.
- A nonprovisional application based on the provisional is filed.

Thus, a provisional application simply acts as a filing date placeholder until a related, nonprovisional application is prepared and filed.

The Background on Provisionals

In June 1995, the USPTO officially started accepting provisional patent applications as a lower-cost method to obtain a first patent filing in the United States.

Provisional Basics

The filing of a provisional application versus a conventional (nonprovisional) application is not an either/or question. You can temporarily protect or reserve a filing date for your invention through the filing of a provisional application. However, within one year after filing your provisional application you will be required to file a nonprovisional application on that same invention or otherwise abandon the application. The filing of a provisional application also acts as the beginning of the one-year clock to file applications internationally to protect your invention overseas (unless you've otherwise given up your ability to file a foreign application, as explained in our discussion of foreign rights in Chapter 9).

One great thing is that the 20-year term of your patent will date back only to the date of the filing of your nonprovisional application, not your provisional application.

The Benefits of Provisionals

The provisional application filing date acts as the priority date for your patent application. After you have filed a provisional application for patent, you can also commence using the expression "Patent Pending" regarding your invention. There are other benefits of provisionals that can be summed up as follows:

- **Low cost.** At the time of writing this book, the government filing fee for a provisional application is $80 for a small entity and $160 for a large entity.
- **Quick development.** The specific requirements for a provisional application are outlined in the following section, but these reduced requirements make for a much quicker application development period versus the comparative development time for a nonprovisional application.
- **Immediate commercial launch.** Filing a provisional application is often used as an "emergency" effort to preserve a filing date when an inventor knows that he or she is going to publicly disclose the invention but there is not ample time to prepare a conventional patent application. This is also the case when a company is getting ready to commercially launch a product but has not filed conventional patent applications in an effort to protect its technology.

- **Confidentiality of provisionals.** All provisional applications for patent received at the USPTO are kept secret. However, if a conventional patent that claims the priority of the provisional application is filed and issues as a patent, then the provisional application becomes a public document.

Provisional Requirements

The provisional application doesn't require filing a formal patent claim, oath, or declaration, or any indication of prior art (an information disclosure statement).

Now that you know what is not required, what is required? Here's the basic list of requirements:

- A specification that would otherwise satisfy the specification requirements for a nonprovisional application (without the claims).
- A description of how the invention is best assembled or replicated.
- Any drawings necessary to understand the invention, just as would be required for a nonprovisional application.
- An identification of the inventors who contributed to the subject matter disclosed in the application.
- The filing fee that we mentioned earlier in this chapter along with the Fee Transmittal Sheet described in Chapter 4.

- A cover sheet. Although no formal cover sheet is required, you can use a cover sheet provided by the USPTO. You can download this cover sheet from the USPTO website at www.uspto.gov/web/forms/sb0016.pdf.

The cover sheet should indicate the following information: The application is a provisional application, the name or names of the inventor or inventors, the residence of each inventor, the title of the invention, the name and registration number of the attorney or agent if any, and the docket number (if any) used by the applicant.

Provisional Pitfalls

As we mentioned, to have a nonprovisional application reap the benefit of the filing date of a provisional application, the nonprovisional application must be filed within one year from the provisional filing date. All provisional applications are automatically abandoned one year after they are filed. Thus, if the nonprovisional application is not filed before the abandonment of the provisional application, the priority date is lost forever.

When preparing a nonprovisional application that is based on a provisional application, the claims that are included in the nonprovisional application must be supported by the information provided in the provisional application. Otherwise, the nonprovisional patent application will not be able to claim the benefit of the earlier filing date of the provisional application.

The largest risk associated with provisionals is just that—the provisional was really insufficient to cover all of the cool features of your invention when it comes time to file the nonprovisional. This can happen very easily, and you should be keenly aware of this risk when deciding to file a provisional application rather than a nonprovisional application. For example, a provisional application may not have a sufficient disclosure for any of the following reasons:

- **Narrow disclosure.** Some provisional applications are simply a copy of a software program, functional specification, or user manual for a product. Such a provisional application will generally support the version of the product that you disclose but, it will not be sufficient to support claims that are directed toward potential alternate designs or improvements.
- **Time constraints.** Many times, companies file provisional applications just prior to publicly disclosing their inventions at trade shows, through press releases, or putting products on the market. Time constraints often lead to inadequate disclosures that cannot support a comprehensive nonprovisional application.
- **Invention shifting.** Let's face it—what you thought you were going to do might change. As a result, the provisional application that you filed last year may now be insufficient to cover the invention as it exists today.

As we mentioned, these pitfalls have undermined many well-intentioned individuals who filed provisional applications. As a result, we highly recommend that you revisit your provisional application well before the annual anniversary when you must file a nonprovisional application. Early review will help you determine the extent to which you're able to rely on the provisional application in creating the nonprovisional.

On a final and related pitfall point, it's important not to wait until 11½ months after you file a provisional application to speak with an attorney about filing a conventional application or to begin preparing a conventional application on your own. Some people believe it's an easy and quick process to convert their application. In fact, it's not. Obviously, the best approach is to start working on the nonprovisional application promptly after filing your provisional application. However, at the very least, you should start to work on the nonprovisional application three to four months prior to the one-year anniversary of the filing of the provisional.

Good Counsel

You can't file a provisional application for design patents.

Do You Need an Attorney?

The USPTO doesn't require you to use an attorney to file a provisional patent application. However, you're not released from any of the requirements we mentioned for a provisional application as a result of not using an attorney. Therefore, you need to consider whether you should use an attorney and precisely what you'll request from her or him. The one thing to keep in mind here is that using an attorney to develop and file your provisional application may significantly increase the intended low-cost benefit you were trying to achieve with a provisional patent application.

The most important thing you should consider when confronting the attorney question is whether you believe you'll be able to write a sufficient specification for the provisional. Remember, the attorney will need to draft a claim set and will be limited by what you've written in the specification.

An approach many people take is asking their attorney to read through the written specification once and provide comments that they can incorporate. This *once over* approach can prove very helpful at addressing this primary concern while maintaining a relatively low cost structure. However, beware, many patent attorneys do not like this approach because, invariably, many clients will complain a year later when certain critical elements of their invention are not included within the specification and important patent claims cannot be supported as a result.

Where to Mail Your Provisional Application

After you have completed the requirements outlined earlier in this chapter and provided the information needed on the cover page, mail the cover page, provisional application, return receipt postcard (described in Chapter 3) and appropriate check to:

Mail Stop Provisional Patent Application
Commissioner for Patents
PO Box 1450
Alexandria, VA 22313-1450

As a reminder, the cost to file a provisional application is $160 for large entities and $80 for small entities. The application should be sent via Express Mail so that the date that you mail the application will be your official filing date. Make sure you keep a file containing this mail receipt and a copy of each of the application, filing forms, and check sent.

Other Preliminary Filings

Other than filing a provisional application for patent or a nonprovisional patent application, the USPTO also allows you to file two other patent-related documents: a Statutory Invention Registration (SIR) and an Invention Disclosure Document (IDD).

- A Statutory Invention Registration enables you to publicly disclose an invention so that no one can claim ownership over a particular invention.

- An Invention Disclosure Document establishes that you came up with an idea at least as early as the filing date.

The Least You Need to Know

- A provisional application can provide a lower-cost, preliminary step to a conventional or nonprovisional application.
- You have one year to file a nonprovisional application after filing a provisional application.
- You need to make sure your provisional application adequately describes your invention or you might not gain the benefit of your filing date or, worse, you might miss out on a patent opportunity altogether.

Chapter 7

From Application to Issued Patent

In This Chapter

- The procedures for prosecuting a patent application
- How to respond to an office action from the USPTO
- The reasons an examiner may reject your claims
- Your options in communicating with the examiner

Have you ever watched people play the war board game of *Risk?* We don't mean while you played; we mean totally sitting back and watching with no participation whatsoever. It has got to be just two notches up the excitement scale from watching paint dry or grass grow. If you are the type who does not understand what we are talking about, then the next phase of the patenting process may be just fine and dandy for you.

After you file a patent application, very little happens over a long period of time. The USPTO is absolutely inundated with more work than the employees can handle, and quite a large backlog of applications has built up at the USPTO. So if you are the type of person who demands instant gratification, filing a patent may not be the best experience for you. For utility applications, other than some preliminary communication from the USPTO, your patent will most likely not be examined for two to two and a half years after it was filed. However, when the USPTO picks up the case for examination, things roll relatively quickly.

In this chapter, we describe each of the phases that you will or could go through after filing your patent application with the USPTO.

The Filing Receipt

When your patent application is received at the USPTO, it is initially reviewed for completeness. The USPTO verifies that you have filed all of the necessary documentation, paid the required fees, and that no pages are missing in the application. Upon verifying that the application is in order and ready for examination, the USPTO will mail the applicant a filing receipt.

The filing receipt is a very important document in that it is your first official verification that the patent application has been filed. You should check all of the content of the filing receipt to ensure that it is accurate. The USPTO can easily make mistakes on entering data into their system, and this is your first chance to catch the mistakes.

If you determine that an error exists in your filing receipt, you can prepare and file a request for a corrected filing receipt. When filing the request, you need to submit a copy of the original filing receipt, marked up with a red pen to indicate the errors. If the error is the USPTO's fault, then there is no additional fee. If the error is the applicant's fault, a fee is assessed for the issuance of a new filing receipt.

Notice to File Missing Parts

When the USPTO determines that the patent application is missing items, they issue a Notice to File Missing Parts. This notice generally includes a two-month time frame in which to respond and indicates the deficiencies identified by the USPTO. The typical deficiencies include a failure to pay the necessary filing fees or failing to file a proper oath or declaration.

The USPTO charges a surcharge for the late payment of a filing fee or the filing of an oath or declaration. At the time of writing this book, the surcharge is $130 ($65 if you qualify as a small entity).

When receiving a Notice to File Missing Parts, you must file a proper response within the requested time frame or the application will be considered abandoned. You may obtain extensions of time by paying an additional fee but no extensions of time are available beyond six months from the date that the USPTO mailed the notice.

Assignment to an Examiner

After the application is verified as being complete, it is assigned to an art group for examination and then is placed into a queue for that art group. The patent is then either assigned or picked up by one of the examiners in the art group and tentatively scheduled for examination. The schedule is tentative because other applications under a petition to expedite can arrive and take priority over a previously filed application.

When it is time for the examiner to *prosecute* your case, the examiner first conducts a search for prior art related to the invention disclosed in your patent application. The examiner then reviews the patentability of your invention based on this prior art.

Legal-Ease

The term **prosecute** refers to the negotiation of the patent application between you and the examiner assigned to the application.

When the examiner is finished reviewing your case, the examiner prepares and mails an office action. The office action will set forth any rejections and/or objections for some or all of the claims and possibly objections to the patent application.

Restriction Requirements

The examiner may issue a restriction requirement if he or she determines that your claims are directed toward more than one invention. In this situation, the examiner will identify groups of claims that are related to the same invention and request you to elect examination of the claims in one of the groups. You will have the opportunity to either elect a group of claims or to present arguments to the examiner as to why the restriction requirement is not necessary.

Objections to the Drawings

A draftsperson at the patent office will review the drawings that you submit with the application. The drawings have to be in conformance with the guidelines published by the USPTO and addressed in Chapters 4 and 5 of this book. If your drawings are objected to, the examiner may request you to fix the drawings before further examination can occur, or may allow you to present amendments in the response. In either case, prior to paying the issue fee for the application, you should make sure that the drawings are in proper form and that the USPTO has accepted them as formal drawings.

Objections to the Patent Application

The examiner may reject your application as not providing an enabling description of what is claimed. For example, if the specification does not teach someone how to implement the claim, the examiner can state that it is not enabled.

The examiner may also identify errors or typos that appear in the specification and figures and request you to correct them. In amending the patent application, it is important to remember that you cannot add any *new* subject matter. New subject matter is information that was not included in the application as filed, including the claims, figures, and specification.

Rejections of the Claims

If the examiner believes that one or more of your claims is disclosed in the prior art, the office action will include a rejection based on a lack of novelty. If the examiner believes that one or more of your claims would have been obvious given the prior art, the office action will include a rejection based on obviousness.

Just the Facts

A rejection based on a lack of novelty is referred to as a 102 rejection, and a rejection based on obviousness is referred to as a 103 rejection. These rejections get their names from the law number that defines the rejection: 35 U.S.C. 102 and 35 U.S.C. 103.

Objections to the Claims

The examiner may also object to one or more of your claims for various reasons. A common reason is that the claim contains errors or is not understandable by the examiner in the manner in which it is presented. More detrimental objections are when the examiner states that the claim does not include patentable subject matter or is not supported in the specification or the drawings.

Filing a Response to the Office Action

The applicant is generally allotted three months from the mailing of an office action in which to respond. The applicant can obtain extensions of time for responding to the office action, but the fee increases with each month extension. However, the applicant must file the response no later than six months following the mailing of the office action or else the application will be deemed to have been abandoned.

In responding to the office action, the applicant must address each and every point raised by the examiner. Thus, for each claim that is rejected, the applicant must either amend the claim to overcome the rejection, cancel or withdraw the claim, or present an argument refuting the examiner's basis for rejection—presenting such an argument is referred to as a traversal.

The response to an office action must meet certain formatting requirements. The front page must include the identification of the application by providing the title, serial number, filing date, and the name of the first inventor. In addition, the front page should identify the examiner, the art unit to which the case is assigned, and the attorney docket number.

The title of the response should describe the contents. For example, if the response provides an amendment to the claims, the title should be "Amendment and Response to Official Action." If no amendment is presented, the response should simply be titled "Response to Official Action."

At a minimum, the response should include a claims section and a remarks section. Each of these sections should start on a separate page.

The Claims Section

The claims section should list all of the claims pending in the application along with a status for each claim. The currently accepted status indicators include previously presented, currently amended, canceled, and withdrawn. The text of each claim in the application that is not being canceled must be included in the response. The format for the claim and the status indicator should conform to the following:

> Claim 1. (previously presented) A method for …
>
> Claim 2. (currently amended) The method of claim 1, further comprising the step of ~~temporarily~~ storing data into a memory device in a first in first out queue.
>
> Claim 3. (canceled)

Claim 2 shows an amended claim. Any amendments to the claims must show new text as underlined and deleted text with a strikethrough line. Alternatively, deleted text can be enclosed in square brackets, for example, [passport].

The Remarks Section

The remarks section should address every point raised by the examiner. In addressing each point, the remarks should state the action taken regarding the claim. For example, the remarks could include one or more of the following statements:

- The claim has been amended to overcome the rejection or objection.
- The claim has been canceled or withdrawn so the rejection or objection is not being addressed.
- An argument stating why the rejection or objection is unwarranted should be presented.

Overcoming a Novelty Rejection

For a claim to be rejected based on novelty, the examiner must find one piece of art that discloses each and every element of the rejected claim. To overcome a novelty rejection, you either amend the claim to include an element that is not described in the cited reference or present an argument as to why the examiner is not correct in stating that one or more elements are included in the cited reference. Often you will find that the examiner may refer to a portion of a cited reference and allege that it discloses the element of the claim. However, upon careful review, you may find that the examiner is either flat wrong or may not have a proper understanding of the technology.

In responding to such a situation, you should respectfully describe why the examiner is not correct.

Overcoming an Obviousness Rejection

Generally, the examiner can reject a claim if, by combining two or more references, each of the elements of the rejected claim are described, and the references are related in such a manner that it should have been obvious to combine the references. In addition, the examiner may reject a claim on obviousness if most of the elements are disclosed in one or more references, and the elements that are not disclosed are merely obvious extensions.

In addressing an obviousness rejection, you need to establish why it would not be obvious to combine the references, or why the nondisclosed element is not merely an obvious extension.

Overcoming Claim Objections

Oftentimes, the examiner will object to your claims. The objection can be based on one or more of several issues. One objection is that the claim is not in a proper format or is not understandable as written. The examiner will request you to either cancel such a claim or amend it to overcome the objection.

Another objection is when the examiner determines that a dependent claim is allowable but the independent claims or dependent claims that it depends from are not allowable. In this situation, you will be requested to amend the objected to claim to incorporate each of the limitations of the other claims from which it depends. The language in the office action refers to these claims as the base claim and any intervening claims.

A common objection or rejection basis that appears is that the claim includes an element that does not have a proper antecedent. What this means is that your claim includes an element that has not been properly defined. The following two example toothbrush claims include the element of a handle portion of the toothbrush. The first example includes a proper antecedent basis for the handle, and the second example does not.

> **Claim 1.** An apparatus for cleaning teeth comprising:
>
> a handle portion having a first and a second end;
>
> a plurality of bristles fixedly attached to the first end of the handle portion; and
>
> a rubber covering encasing a particular length of the second end of the handle portion.

> **Claim 2.** An apparatus for cleaning teeth comprising:
>
> a plurality of bristles fixedly attached to the first end of the handle portion; and
>
> a rubber covering encasing a particular length of the second end of the handle portion.

In the second claim, the term "the handle portion" is used without being properly introduced as in the first claim.

Check the Dates on the References

When the examiner cites references in support of a rejection, you should always check the filing date of the references to make sure that they were filed or published earlier than your filing date or your date of invention if it is earlier than your filing date. The references are not valid references otherwise.

If the references are earlier than the date you filed your patent application but later than a clearly documented date of your invention, you can submit an affidavit that *swears behind* the reference. Swearing behind simply means you sign a statement indicating that you invented the subject matter of the claim at a date earlier than the filing date or publication date of the cited reference. You will need to submit evidence along with the affidavit.

An Interview with an Examiner

You also have the opportunity to conduct an interview with the examiner, either on the telephone or by visiting the examiner's office. An interview can go a long way in helping to understand the examiner's position and conveying your arguments to the examiner. The office action will include the examiner's telephone number, and you are free to call the examiner and schedule an interview. You should be prepared for the interview in advance and be ready to clearly articulate your position and ask the examiner questions.

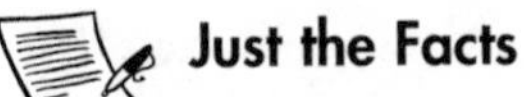

Just the Facts

In corresponding with the examiner, you should capitalize the word "Examiner." This practice is followed by applicants and attorneys because we are required to conduct business with decorum and courtesy. So we capitalize the word "Examiner" out of respect. This requirement can be found in the Code of Federal Regulations, 37 CFR 1.3.

Extensions of Time

We mentioned earlier that you can obtain an extension of time for responding to most communications from the USPTO. An official communication will set a period of time in which you can respond for free. Any delay beyond this period of time will cost you, and you cannot delay beyond six months from the original date the official communication was mailed.

The fees associated with an extension of time can change from year to year.

Final Office Actions

After you have filed a response to an office action, the examiner can issue a final office action if …

- Your claim amendments would require an additional search.

- Your arguments are not persuasive and the rejection remains.
- The examiner rejects the claims on previously cited art.

You can file a response to a final office action. However, in general, the USPTO will only allow you to amend the claims in a manner to put them into condition for allowance.

What Next, After a Final Office Action?

If you received a final office action that includes allowable claims, you can cancel all of the remaining claims and simply request the USPTO to issue a patent based on the allowed claims.

If you received a final office action that does not include any allowable claims, you may have to refile the application along with a Request for Continued Examination (RCE). When filing an RCE, you have to pay another filing fee as well as submit the equivalent of a response that either amends claims, argues against the examiner's rejections, or both.

Rather than filing an RCE, you can also file an appeal. An appeal involves requesting an independent review of the examiner's position. The appeal allows you to submit an appeal brief that presents arguments in support of your claims. Filing an appeal also includes a filing fee. When filing an appeal, you can also request an oral hearing during which you have an opportunity to present your case to the USPTO. Of course, you can just give up.

Notice of Allowance

If you are able to convince the examiner that one or more of your claims should be allowed, and after you have canceled the remaining claims (if necessary), you will receive a notice of allowance. The notice of allowance identifies the claims that are allowed and may include a statement from the examiner as to why the claims are allowed. If the examiner's statement imposes any additional unwarranted limitations on the claims, it is good practice to file a response to refute the limitations.

Usually the notice of allowance is accompanied by a notice that the issue fee is due. Again, you are provided with a time period in which to submit the payment of the issue fee. However, it is important to note that the issue fee due date is listed on the notice and cannot be extended. If you don't pay the issue fee by the provided date, the patent is abandoned.

Important Considerations Before the Patent Issues

Prior to your patent issuing, you need to make another decision. You can file a continuation application or a continuation-in-part application up to the date that your patent issues. You may want to consider filing a continuation application if you have canceled some of the claims during the prosecution of the application. The continuation application can be used to further attempt to get these claims allowed. A continuation-in-part application can be filed if you have added new aspects to your invention. In either case, the application must be filed prior to the issuance of the related application.

Another good practice is to look at the allowed claims in view of your competitors in the market. Ask yourself the question, "Do my claims cover my competitor's products?" If not, you might want to consider either filing a continuation application with claims that would cover your competitor's products or canceling your application from being issued and re-file the application as an RCE with claims amended to cover your competitor's products.

The Least You Need to Know

- You need to review your filing receipt from the USPTO carefully to ensure it is correct.
- You should not always assume the examiner is right; you may need to educate the examiner regarding the technology of your invention.
- You can give up some claims to get an application to issue, but you have the right to file additional patents with those claims at a later time.
- Be careful what you write in responding to an office action; your statements can be used to limit the scope of your claims.

Chapter 8

The Life of a Patent

In This Chapter

- When is a patent application published?
- Can you keep a patent application secret?
- The duration of an issued patent
- Extending the duration of a patent
- Maintaining the patent

The first patent in the United States—Patent No. 1—was issued on July 31, 1790, to Samuel Hopkins, of Pittsford, Vermont, and was directed toward a new and improved apparatus and process for the making of potash and pearl ash. Patent No. 1 was signed by President George Washington, Attorney General Edmund Randolph, and Secretary of State Thomas Jefferson. Patent No. 1 was a great advancement in the process of making potash and earned Mr. Hopkins considerable revenue through licensing. Is Patent No. 1 still valid today? No, Patent No. 1 expired on July 31, 1804, 14 years after it was issued.

As you have already learned, a patent is a monopoly granted by the government to an inventor and is only active for a limited time. Why do patents expire? The overall purpose of our country's patent program is to promote the useful arts and sciences. A limited monopoly allows the developer of new technology to benefit from that technology for a period of time and then the rest of the public can benefit from the technology. If the patent endured forever, then the useful arts and sciences would most likely be stagnated rather than promoted.

In this chapter, you will learn to determine how long a patent will be active and when the clock starts to tick on the life of a patent. In addition, you will learn what needs to be done to ensure that the life of a patent is maximized.

Publication and Issuance

When a patent application is filed, two public events may ultimately occur. The first event is the publication of the patent application. The second event is the issuance of a patent.

Publication of the Patent Application

In late 1999, the United States Congress enacted the American Inventors Protection Act. One of the changes this act introduced into the patenting process is the publication of a pending patent application. The publication of a patent application happens at a very specific time—18 months after the filing date (or the earliest claimed priority date, if the patent is based on a previously filed non-provisional patent or foreign patent and is using the filing date of the previous application as the priority date). Prior to this change, a U.S. patent application was held in secret until a patent was issued.

At the time you file the patent application, you can request the Patent Office not to publish your application. If you don't make this request at the time of filing, you cannot subsequently make a request to stop the publication of the patent application unless you abandon the application. Filing a nonpublication request can be accomplished by completing a simple form that can be downloaded from the United States Patent and Trademark Office (USPTO) website at www.uspto.gov. The form is labeled as PTO/SB/35.

If you file a nonpublication request, you can rescind the request at any time. Once rescinded, the application will be scheduled for publication 18 months from the filing date. If you subsequently file the patent application with the Patent Cooperation Treaty or into a foreign country that requires publication at the 18 month date, you must notify the USPTO regarding this filing within 45 days. If you fail to notify the USPTO, your application will be abandoned.

To Publish or Not To Publish

There are trade-offs that must be considered in determining whether to have your patent application published at the 18-month date. If you initially filed your patent application with a nonpublication request, and later you determine that someone is potentially infringing one or more claims, you should rescind the request and put the potentially infringing party on notice.

Good Counsel

You can view published patent applications by visiting the USPTO website at www.uspto.gov. Once a patent issues, it is removed from the list of applications and is moved to the list of patents that can be viewed at the USPTO website.

Issuance of a Patent

If you are able to successfully convince the USPTO that your invention as claimed is patentable, the USPTO will issue a notice of allowance with issue fee due. After you pay the issue fee, the ball will start rolling on assigning a patent number and publishing the allowed patent. When the patent is issued, your monopolistic reign begins and will continue until the patent expires.

The Span of Protection

The span of protection for a patent can depend on when the patent application was filed or when the patent issued. The duration of protection has changed throughout history.

Patents in the Past

In the Patent Act of 1790, the term of a patent was set at 14 years from the date of issuing the patent. In 1861, the term of patent protection was extended from 14 years to 17 years from the date of granting the patent. Basing the expiration of a patent on the issue date turned out to be problematic due to a concept called *submarine patenting*. Submarine patenting is when an inventor purposely keeps a patent pending in the USPTO for several years (hiding like a submarine under water) while waiting for the current state of the technology to catch up with the invention. Once the invention becomes commonplace, the inventor then allows the patent to issue (fires the torpedo). The inventor then gets to sit back and enjoy 17 years of monopolistic power.

Today's Patent Terms

For any patent applications filed after June 8, 1995, the term of patent protection is 20 years from the filing date or the earliest claimed priority date. For a patent filed on or before June 8, 1995, you have to make two calculations to determine the patent term. The first calculation is adding 20 years to the filing date. The second calculation is adding 17 years to the date that the patent issues. So what is the term for such a patent? You get to pick whichever of the two calculations gives the longest term.

If your patent was based on a provisional application, the earliest priority date will either be the filing date for the nonprovisional patent or, if the application is based on a foreign or PCT application, the filing date of such patent application.

Up Periscope

The new laws that require the term of a patent to be calculated based on the filing date rather than the issue date have effectively put an end to the reign of submarine patents. Because the term of protection is tied to the filing date rather than the issue or grant date, it is in the best interest of the patent owner to have the patent issue as soon as possible.

The Term of a Plant Patent

The term of protection for a plant patent is identical to that for a utility patent.

The Term of a Design Patent

The term of protection for a design patent is still tied to the issue date of the patent. Currently, the term for a design patent is 14 years from the date of grant.

Adjustments to the Patent Term

Patent rights are not enforceable until a patent is issued. And with the term of a patent being tied to the filing date, delays in the prosecution can greatly impact the rights of the patent owner. If the delays in prosecution are due to the USPTO, this impact is not fair for the patent owner. To address this concern, the patent laws were changed to allow adjustments to be made in the term of the patent application.

The adjustments available depend on the filing date of the patent application. No adjustments are available for patents filed on or before June 8, 1995. For patents filed after June 8, 1995, but before May 29, 2000, one set of adjustment rules are applied; a different set of adjustment rules are applied for patents filed on or after May 29, 2000. For more information, go to www.uspto.gov/web/offices/dcom/olia/aipa/infoexch.htm.

Good Counsel

Does the USPTO lose patent files? Yes, they certainly do, and you should be aware of it. In fact, you should follow up with the USPTO every six months to make sure they know where your file is.

Good Counsel

To maximize the term of your patent, you should be diligent to respond to requests from the USPTO within the time frame that is requested.

Determining the Adjustment Term

So who does all the math work to determine whether the term of your patent should be extended? The USPTO does! The USPTO will determine the period of adjustment and include a notice of that determination in the written notice of allowance. But the smart thing to do is to always check their math.

After you receive the notice of allowance, take out your file and your calculator and do some double-checking. You do have the opportunity to request a change in the term adjustment, so it could be very beneficial to go through the file history and add up any delays by the USPTO and then subtract any delays caused by you or your attorney.

In addition, if the USPTO offsets the extension of time due to their delays because you also caused some delays, you may be able to capture some of that time back. For example, if you can demonstrate that in spite of your diligence to respond within the allotted time frames, you ended up causing a delay, you can request the USPTO to not consider those delays. You must make such a request prior to the issuance of the patent. In this situation, it doesn't hurt to ask!

Appeal

If you disagree with the adjustments to the term of your patent, or the lack of adjustments, you still have one more remedy. You can initiate a civil action against the director of the USPTO. That sounds a little scary, I know, but it really is not that scary—or personal. For example, it is not going to hurt the director's feelings, and he is not going to be called into the president's Oval Office for a good scolding.

Any appeal regarding the adjustments to the patent term should be filed in the U.S. District Court within 180 days after the patent has issued. Your appeal simply sets forth the reasons you disagree with the adjustments to the patent term. If the court finds in your favor, the director will be instructed to alter the term of the patent to reflect any changes granted by the court.

Renewals and Fees

So what happens after the patent issues? Do you just sit back and wait out the term of the patent without having to deal with the USPTO anymore? Well, close but no cigar. The USPTO requires you to touch base periodically and, when you touch base, they make you leave a little cash behind. These cash deposits are called maintenance fees. Maintenance fees are only required for utility patents. So if you have a design patent or a plant patent, you get off easy.

The USPTO requires you to pay three maintenance fees during the term of your utility patent. Each maintenance fee is due and payable within a six-month window. The windows of time for the payment of the maintenance fees are calculated based on the date the patent was issued. The following table lists the payment windows.

Payment Windows for Patent Fees

Maintenance Fee	Window from Issue Date
First	3 to 3.5 years
Second	7 to 7.5 years
Third	11 to 11.5 years

What happens if you miss paying the maintenance fee within the six-month window? Well, the USPTO does give you another chance, but it will cost you a little bit extra. You can actually pay the maintenance fee up to six months after the end of the respective window (the six-month grace period) by paying a surcharge fee.

The maintenance fees and the penalty fee can change from time to time, and the current fee schedule for the USPTO should be checked prior to making a payment. The following table lists the current fees (at the time of writing this book).

Maintenance and Penalty Fees

Description of Fee	Fee Amount Large Entity	Fee Amount Small Entity
First	$ 910	$ 455
Second	$ 2,090	$ 1,045
Third	$ 3,220	$ 1,610
Surcharge	$ 130	$ 65

What happens if you don't make the payment by the end of the six-month grace period? Basically, the patent is abandoned at this point. However, you still have the option to file a petition based on the missed payment being unintentional or unavoidable. If the USPTO buys into your reasons, then the patent can be reinstated.

How to Pay Your Maintenance Fees

The best way to pay your maintenance fees is electronically over the Internet at www.uspto.gov. The USPTO has a nice web utility—once you enter the serial number for the patent, the fee that is due appears in your browser window. You can then enter your credit card number and be done with it. If you would rather mail your maintenance fee payment, you can send the fee payment to the following address:

> United States Patent and Trademark Office
> PO Box 371611
> Pittsburgh, PA 15250-1611

When paying the maintenance fee by mail, you should use the Fee Transmittal form, a return receipt postcard, and use Express Mail, all of which is described in Chapter 5.

Expiration or Termination

The patent can either expire naturally at the end of the patent term, or it can be terminated earlier by your failing to pay the maintenance fee or by having the patent invalidated in a court proceeding or denied patentability during a reexamination proceeding. What happens when the patent expires or is terminated? Well, the good news is that it will not cost you any money for the patent to expire or terminate. However, upon the expiration or termination, your rights to exclude others from making, using, and selling your invention also die with the patent.

The Least You Need to Know

- A patent application will become a public document 18 months after it is filed, unless you request the USPTO not to publish the patent application.
- Today, a patent will be active for 20 years from the date the patent application was filed.
- The term of a patent can be extended if the USPTO delays in processing your patent application.
- After a utility patent is issued, you still need to pay maintenance fees to keep the patent active.

Chapter 9

Patents on Foreign Turf

In This Chapter

- Filing international patents
- Overview of the Patent Cooperation Treaty
- Filing a PCT application

As a young boy living in the Republic of Panama, one of the co-authors of this book, Gregory, had the opportunity to take Spanish beginning as early as the third grade. Being a little creative, Gregory would oftentimes try to piece together what he was learning in class in an attempt to communicate with the locals. On one occasion, his family was shopping in a small village and Gregory wanted to know the price on certain items being sold. Gregory remembered in class that the word *como* meant "how" (although actually it meant "as"), and *mucho* meant "much." So Gregory put them together and would lift an item in the air and inquire, "*¿Como mucho?*" Gregory got plenty of smiles, puzzled looks, and attention, but no cost quotations. Upon returning home, Gregory asked their native language–speaking maid what the problem was. When he told her that he would ask "*¿como mucho?*" she started laughing out of control. When she finally regained her composure, she informed him that in the native dialect, he was telling everyone that they "eat too much."

The good news about filing patents internationally is that the language barrier is not so much of an issue. There are mechanisms set up to ease the pain of filing applications in other countries, and it is quite easy to solicit the help of a foreign associate to aid in prosecuting a patent in their country.

In this chapter, we describe the ways in which you can file a patent application in a foreign country. We will also discuss certain treaties that allow you file a single patent application that can qualify for being filed later in a large number of countries.

International Patents

A patent in the United States enables you to prevent others from making, using, selling, offering to sell, or importing items within the boundaries of the United States. However, your U.S. patent cannot be used to prevent someone from making your product in Mexico and selling it in Europe. To obtain protection in another country, you have to have a patent in that country.

You should also be aware that the patent laws in other countries can differ from the laws in the United States. One of the biggest differences is that most foreign countries require *absolute novelty*.

Absolute novelty means that if the invention is published prior to the filing of a patent application, the inventor cannot then file the patent application. In the United States, inventors are granted a grace period of one year in which to file a patent application after the technology is made public.

The cost associated with obtaining a patent in various countries can vary greatly in amounts due and the timing of payments. Another important consideration is that some countries require you to manufacture a patented product within their country within a certain period of time. Depending on the country, failure to do so may result in voiding the patent or providing a compulsory license to others desiring to manufacture the product.

Oftentimes you will hear people refer to a *worldwide patent*. Technically speaking, there is no such thing. There is no one single place where you can file a patent that will grant you worldwide patent rights. There are ways to file a single application that *can* ultimately be filed in other countries. However, each country in which you obtain patent protection ultimately has to issue you a patent.

Just the Facts

Because the deadline to file a foreign application claiming priority of a U.S. application is 12 months from the date of filing the U.S. application, you may have to file your foreign applications before you know if the U.S. patent will issue.

The Patent Cooperation Treaty

The Patent Cooperation Treaty, also known as the PCT, is an international agreement that facilitates obtaining international patents. More than 100 countries are members of the PCT. The PCT operates as a central depository for filing patent applications. An application filed with one of the receiving offices in one of the PCT member countries can subsequently be filed in any or all of the countries that are members of the PCT. In addition, if you file a PCT application within one year of filing a national patent application, you can claim the priority date of the national application as the filing date for the PCT application.

Procedures for Filing a PCT Application

You can file a PCT application up to one year after filing a U.S. patent application (thanks to the Paris Convention). When filing a PCT application, you must complete the proper application forms. These forms are very different from the forms required for a U.S. patent. You can download the PCT filing forms at www.uspto.gov/web/offices/pac/dapps/pct/chapter1.htm.

At a minimum, you will need to download and complete the Request form PCT/RO/101 and, if you are filing the application with the U.S. receiving office, you will need to complete the Transmittal Letter to the U.S. Receiving Office form PTO-1382. These forms need to be filed along with a complete copy of the patent application.

Legal-Ease

The **priority date** is the date of filing your patent application. The priority date can be based on a U.S. provisional or non-provisional filing, filing in a foreign country, or filing with the PCT. The earliest filing date is your priority date. This date can be used for other filings as long as you file within one year of your priority date.

You can either file a PCT application prior to filing any other applications or you can file the PCT application after filing a national application in one of the PCT member countries. If you file the PCT application after filing a national application, you will have to provide the requested information relating to the national application. In addition, depending on where you filed the national application, you may need to obtain a certified copy of the application and provide that to the PCT receiving office. If you filed the original application in the USPTO and the PCT application in the U.S. receiving office, then you do not need to obtain a certified copy.

The PCT Search Report

Once the PCT application is filed, the receiving office with which you filed the PCT application will conduct an initial search to determine patentability of the invention. The results of the search will be reported to you along with the examining authority's view regarding the relevance of the search results to the claims. At this point, you are offered an opportunity to amend the claims prior to the application being evaluated.

Preliminary Examination of the PCT Application

Shortly after providing the search report, the PCT application can enter into Chapter II, which is when the application is evaluated by the receiving office—the preliminary examination. The PCT application is evaluated for patentability pursuant to the PCT guidelines and based on the results of the search. Upon completion of the evaluation, the evaluating office issues a preliminary examination report providing an opinion regarding the patentability of the invention. The patentability opinion will simply state whether the claims are novel, non-obvious, and include patentable subject matter.

Going National

Thirty months from the filing of the PCT application, or the priority date of the national application that the PCT application is based on, you must go into the national phase. During this phase, you need to officially file the PCT application with the national patent office of each country in which you want a patent to issue. This process can be rather costly because each country will have its own filing fees, and in some countries you will be required to pay for a translation of the patent application into the native language. There are more than 100 countries that are members of the PCT. You can find the list of member countries at www.uspto.gov/web/offices/pac/dapp/pctstate.html.

Benefits of a PCT Application

One of the main benefits of filing a PCT application is that, for a fee, you can delay making a decision regarding national filing for several months. During this period of delay, you will hopefully receive some feedback from the USPTO (if your PCT application is based on a U.S. application) regarding the patentability of your invention. If things look favorable, then you have not lost the opportunity to file your patent in other countries If things look bleak, then you can refrain from filing the patent application in other countries and, thus, save yourself some money.

Another benefit of filing a PCT application is that if you receive a favorable report, you will have a better idea as to how the application will be treated in the other countries. Although the PCT preliminary examination report is not binding on other countries, it can be very persuasive. In addition, at a minimum, filing the PCT application may greatly reduce the amount of prosecuting that has to occur in the various countries.

A disadvantage in filing a PCT application is that at 18 months from the priority date, the application will be published. If you only file a U.S. patent application, you have the option to keep your application secret until it issues as a patent. Filing the PCT application removes this option. In addition, if you filed a U.S. patent application and requested for it not to be published, then after filing a PCT application, you have to rescind your request not to publish for the U.S. patent application. Otherwise, the U.S. patent application will be abandoned.

International Strategies for Filing Patents

Filing a patent application in other countries, as we mentioned, can be quite costly. It is not a practice you should enter into without much thought and consideration. Individuals or small companies have a whole different set of thought processes in regard to international strategies than large international companies do.

The general questions you need to ask yourself regarding filing a patent application in another country are whether or not there is a market for your product in that country, whether it makes sense from a cost perspective to manufacture your product in that country, and whether you believe others from that country will be or are competing with you.

Other considerations you need to weigh are whether you will have a presence in the country; for example, will you be able to determine if anyone is infringing and enforce your rights? How well are patents held by foreign entities treated in that country? Do the courts tend to hold them invalid more frequently than patents held by natives of the country? What percentage of your revenue will be coming from that country versus other countries?

Just the Facts

The cost of filing a PCT application increases if your application is longer than 30 pages. Thus, it is good practice to be sure you format the document and take full advantage of the margins allowed.

The Least You Need to Know

- You can file one patent application with the PCT and later have that application examined by selected PCT member countries.
- If you file a PCT application based on a U.S. or foreign patent application, the PCT must be filed within one year of filing the U.S. or foreign patent application.
- The cost of filing foreign patents can be quite extensive; you should carefully weigh the benefits of filing foreign patents.
- For a PCT application, you will need to determine in which countries you want to file national applications within 30 months after your priority date.

Chapter 10

Do You Own a Patent?

In This Chapter

- Employee's versus employer's rights
- Patent rights of independent contractors
- How is the ownership in a patent assigned to others?

So you're a really creative employee and you start coming up with all sorts of ideas to improve your employer's business and ways the company could actually implement these ideas. Some of the ideas are related to your job, but others are just great ideas for other departments or the business in general. Also, let's assume that many of these ideas are patentable and your employer works with you to file many patent applications. Who owns these applications? Do you? Should you?

Determination of ownership of a patent can range from simple—when you came up with an invention and file it on your own behalf—to difficult, when two companies are jointly developing a product using independent contractors.

This chapter focuses on making ownership determination much more obvious. We'll cover the ins and outs of employee and employer patent ownership as well as the rights of independent contractors to the inventions they develop while on a job. Finally, we'll also be sure you know how to protect yourself from unscrupulous individuals who may want to steal your ideas through use of a nondisclosure agreement.

The Easy Side of Ownership

Assume you're working for yourself and you come up with a wonderful new invention. You consult an attorney, write up an application, and file it. A few years later the patent issues. Who owns it? Well, this example is an easy one—you do. You were the sole inventor, you did the invention for your own benefit, and you hired and paid for an attorney to draft the application. Unfortunately, not all examples are quite so easy. Inventors oftentimes work for companies, and these companies expect to own any intellectual property created by the inventor.

The Inventive Employee or Contractor

Most companies require their employees and contractors to sign an agreement upon the commencement of their services that clearly indicates that the company owns any inventions that are created in connection with these services. As a result, if an employee or contractor has created intellectual property while doing his or her job, then that intellectual property is owned by the hiring company.

The important thing to note here is that most agreements specifically limit what's owned by the company to inventions that are related to the services or are developed using resources of the company (materials, the company computer, and so on). Therefore, if an employee or contractor develops an invention unrelated to his or her services, the invention is possibly owned by the employee or contractor.

The Intellectual Property Assignment Provision

Most employment agreements and independent contractor agreements have a provision that assigns intellectual property created by that person over to the employer (or company hiring the contractor).

The assignment provision relates to all types of intellectual property that could be developed by an employee—patents, trademarks, or copyrights. With this language, the employer can feel comfortable that anything created by the employee in connection with his or her job will be owned by the employer. A similar position can be easily used with an independent contractor as well.

Good Counsel

If you are requested to sign an employee or contractor agreement, you need to realize that you may be forfeiting some of your rights. We strongly recommend hiring an attorney to review such an agreement and advise you regarding the ramifications of signing it.

Inventions Without an Agreement on IP

In the absence of written agreements, the question of whether a company owns the right to inventions created or developed by individuals who work for that company can be confusing. The answer depends on many factors, including whether the inventive individual is an employee or someone hired on a contract basis (known as an independent contractor). In the following sections, we look at the various rules that impact our analysis.

Hired to Invent Doctrine

Under these rules, if an *employee* is hired to solve a particular problem and an invention results to solve the problem, that invention is the property of the employer. A few factors that are taken into consideration in determining if this doctrine applies include whether …

- The inventor has previously assigned patents to the employer.
- The employer has a practice of assigning patents by similarly situated employees.
- The invention was made during the term of employment.
- The employer originally presented the problem solved by the invention.
- The inventor has authority to direct activities of other employees of the employer.

- The invention is of significant importance to the business of the employer.
- The employer has taken consistent positions previously.
- The employer has agreed to pay royalties to the inventor.
- The employer pays patent procurement expenses.
- The employer was interested in the invention when it was first made by the employee.

These considerations help gauge the likelihood of whether the hired-to-invent doctrine applies. All of the questions do not need to be answered *yes* (or affirmatively) for the doctrine to apply, but if several of them are not true, then the employer may have an issue asserting ownership.

Fiduciary Inventor Doctrine

If the inventive employee is acting as a *fiduciary* for the employer, such as an officer of the company, then the acts of the employee are deemed to be on behalf of and for the benefit of the employer. As a result, it is presumed that the invention the employee developed was intended to be for the benefit of the employer, and the employer will own the invention.

Legal-Ease

A **fiduciary** is an agent of a principal or a company director who stands in a special relation of trust, confidence, or responsibility in certain obligations to others.

The General Rule About Contractors

What happens when an independent contractor invents something and there's no agreement? The general rule is that the invention is owned by the independent contractor! This means you can actually pay a person to create something for you but still not own it. Most people's assumption is that if you pay for something, you own it, but in this instance that's not true. Be sure you have a written agreement in place!

Shop Right Doctrine

Despite the dire tone of the last paragraph and even the uncertainty around the hired to invent doctrine, there is still some hope for the hiring company. If the previously described doctrines do not apply, the company may still benefit from a royalty-free, nontransferable *license* to use the invention under the *shop right* doctrine. Shop rights may arise in an employee relationship and possibly in an independent contractor relationship. This doctrine applies when the company provided the resources to make the invention. Therefore, working on company time, using the company's tools, facilities, or other employees supports the assumption of a shop right. Please recognize, though, that we're only talking about a license and not ownership of the invention in these situations. The inventor can similarly license the same invention to one of your competitors.

Are You an Employee?

Under the previous rules, employers tend to have the advantage in owning an invention even in the absence of an employment agreement. As a result, it is important to understand whether you're considered an employee or not. For many folks, this is a pretty clear-cut question. However, for others in part-time and contract work, the answer is a bit unclear. Various tests have been developed over time to help courts and agencies determine whether someone should be considered an employee or not. If this is an issue for you or your company (because a nontraditional relationship exists between an individual and a company), then you should consult an attorney to learn more about these tests.

Differentiating Ownership from Right to Exploit

Remember, just because you may own a patent doesn't mean you have the exclusive *right* to the invention. There may be situations in which you need to obtain rights to another patent that applies to your very own invention. For example, you may have invented a new type of windshield wiper for an automobile. However, you may need to obtain rights to produce the automobile that the windshield wipers are attached to.

Joint Ownership

Sometimes two individuals get together and invent something. Perhaps two companies enter into a relationship to develop new technologies. A joint owner may freely make, use, sell, and import the invention without getting consent from the other owners. In fact, that owner can even grant a license to any other person to exercise the same rights. The best way to solve this issue is to have a written agreement between the various owners that limits each owner's rights to exploit the patent in these ways. In the absence of a written agreement, it's hard for any one owner to really enjoy one of the primary benefits of a patent—the right to exclude others from using, making, selling, or importing the invention.

Take the following example: In the interest of the Atkins Diet, Kentucky Fried Chicken and Coca-Cola decide to pool their research and development teams to come up with a new product—liquid carbonated meat. Let's also assume that all the employees on each company's teams have signed an agreement assigning their rights in any invention to their respective employers. If the two companies do not create a joint venture to develop the product, but instead maintain their individual status with relationship to the product, problems may arise when it comes time to determine who owns the invention. In the absence of an agreement, both companies have an unfettered right to the invention—probably not the result they intended.

Good Counsel

If a patent is jointly owned, absent an agreement between the parties, each owner is free to license and assign his portion of the ownership. This could be detrimental if a portion of the ownership is assigned to a competitor. You should be very careful when deciding if your patent should be jointly owned.

Selling and Assigning Your Patent

Can you sell your patent? Absolutely! In fact, this happens all the time, and the transfer of intellectual property rights is becoming a very big business. The sale of a patent is actually called a *sale and assignment* of that patent because money is exchanged and there is a formal requirement that the sale of the patent be recorded as an assignment at the patent office.

Assignment Requirements

An assignment of ownership in a patent must be evidenced in a written agreement, and it must be signed by the owner assigning the patent. Thus, an assignment of a patent cannot be performed as an oral agreement if you intend to record the assignment with the patent office (see the following section).

What Gets Assigned?

When you assign your ownership in a patent, the receiving party owns the patent and now has the right to exclude others from making, using, selling, or importing products covered by the claims of the patent. Therefore, if you intend to continue in a line of business that could infringe on the patent, you need to be sure you obtain a license-back of the patent to the extent required to run your business.

Recordation of Assignment

Any assignment of a patent should be recorded in the USPTO. The failure to record an assignment can be detrimental to the company receiving the patent. If an assignment is not properly recorded after being signed, then a subsequent purchaser of the patent may be able to assert certain rights with respect to the patent. Therefore, get your assignments in as early as possible.

To record an assignment, the assignment must be filed with the USPTO. The assignment is basically a contract or agreement that typically identifies information about the patent, such as the patent number and the title of the patent; the person or entity assigning the patent and the person or entity receiving the patent; and the rights that are being assigned.

Typically, the assignments are generated by the purchaser, who gets them signed and files them with the USPTO to ensure that this task is actually done. When submitting an assignment for recordation, the appropriate fee must also be paid.

Each document filed with the USPTO for recordation must include a cover sheet. You can download the cover sheet from the USPTO website at www.uspto.gov/web/forms/.

Good Counsel

When you are purchasing a business from someone else, it is very important to be sure you wrap up the intellectual property rights, including patents ownership. Be sure the agreement provides you with all rights in and to the patents, any pending patents, and any patents that are extensions of the patent.

The Infamous NDA

An NDA is a nondisclosure agreement. An NDA restricts a person's ability to freely discuss and use information disclosed to that person under the NDA. If you plan on disclosing information that you believe is valuable to you and could be damaging if it were publicly known, then you should be sure you have an NDA in place with the person you're disclosing the information to.

Also, remember that the time period in which you need to file a patent application starts when you publicly disclose information related to the invention. If you start disclosing information before you file a patent application, you will lose your ability to obtain foreign patent rights to the invention. A way to avoid this situation is to have an NDA in place that prohibits public disclosure and, thus, avoids the commencement of the one-year filing period.

You can obtain an NDA from an attorney. Because states differ in what they believe are appropriate provisions in an NDA, you should be sure your attorney is well versed in your state's law regarding NDAs.

The Least You Need to Know

- The inventor is the owner of a patent unless it's assigned to another party.
- Ownership of a patent can get complicated when you're dealing with independent contractors and employees in the absence of a signed, written agreement.
- You can sell and assign your patent to another party.
- You should be sure you have an NDA in place before disclosing your invention to a person or company.

Chapter 11

Making Money Off of Patents

In This Chapter

- Licensing your valuable patents to others
- The value or worth of a patent
- Licensing your patent for profit

Patent licensing is a growing marketplace, reaping great rewards for both sides of the transaction. However, in these situations, you must take steps to ensure you're not being unduly disadvantaged. These pitfalls may include simply missing out on a few bucks or providing an exclusive license to a company that never earns you a dime.

In this chapter, we help you understand the basics of patent licensing from both sides of the negotiating table and provide you with some jewels of information that will make the other side sit up and take notice.

Licensing of Patents

The expression *intellectual property* contains the word *property*. As a result, you can treat intellectual property just like any other property. You can rent it (called a *license*), or, as you learned earlier, sell it (called an *assignment*). If you're going to license your patent, you need to decide under what terms you will license it. Or you may be considering licensing a patent from another company, which raises other issues. We'll cover both situations next.

Some companies do not believe in licensing their patents. They view the technology they've created as a competitive advantage and do not want to see that advantage lost. If you find yourself thinking this way, it's perfectly understandable. However, there are many reasons why it might make perfect sense to license your patent, such as …

- You don't have the time, money, or expertise to really develop a marketplace for your invention. Therefore, some percentage of something is better than 100 percent of nothing!
- You may not be able to cover the entire marketplace with your invention. If you're only able to offer your invention in specific geographic locations or only with respect to one product line, then it may make sense to allow other companies to cover the areas you're not able to.
- Your competitors may be able to find another way to accomplish a similar result

than by way of your patented invention. As a result, wouldn't it make sense to have them adopt your method and make money for you? It's a great marketing benefit to say that your competitors are using your patented technology.

- The financial benefit may be very worthwhile. As the saying goes, everything's for sale; it's just a matter of how much someone is willing to pay!

Licensing Your Patent

For purposes of this section, let's consider this example: You thought up a new voicemail feature for phone systems called VoiceAlert. You read this book and took the steps to obtain an issued U.S. patent for VoiceAlert. You're now interested in licensing this feature to telecommunications companies all over the country, and one of them is interested! The company is considering offering your invention as a new feature to its customer base, for which they'll charge a fee. However, you're wondering about the things you should cover in the patent license agreement.

- **License.** You are entering into a licensing relationship between you (the *licensor*) and the company licensing your invention (the *licensee*). As you know, as a patent owner you have the exclusive right to exclude others from using, selling, making, and importing your invention (as well as a few other things). Therefore, you are granting the licensee a license to do something you could ordinarily exclude them from doing. In this case, they'll certainly need the right to "use" the invention. Perhaps they'll need to manufacture equipment with your invention included so they'll need the right to make the invention as well as import it from their foreign manufacturing facility. These are just a few examples that you'll need to think about. Our advice: Be sure you limit the licensee's rights to exactly what they'll be doing.

- **Money.** You can request some payment up front for the right to license your patent, or you may just want to make money each time a product is sold with your patented feature incorporated. Perhaps you can negotiate for both. Typically, this is a risk-reward analysis. You could probably make more money from sale royalties, assuming that the licensee is successful in marketing the voicemail feature. However, if you prefer safe money—otherwise known as *up-front money*—you'll probably settle for less (but at least you know you'll get something!). You should also request the right to audit the books and records of the licensee to ensure that they're reporting the licensing revenue to you accurately. If your license includes royalty payments, you should also negotiate a minimum royalty payment, which, if not satisfied, allows you to terminate the license.

- **Exclusivity.** Will anyone else be permitted to license the patent? You need to decide whether the license will be exclusive to the licensee or not. If you do decide to make the license exclusive, this means you're not allowed to license the invention to anyone else. If this is the case, you should consider whether the exclusivity is tied to any performance requirements (such as the sale of goods). Just imagine a situation in which you give someone an exclusive license for the life of your patent and they're supposed to pay you $1 for each product sold. What if they don't sell any products? You make nothing! This may sound absurd, but it happens all the time.
- **Business model.** What happens if the licensee sells the VoiceAlert feature as part of an overall package of features to a customer? What percentage of the purchase price should be allocated to your invention that you receive royalties on? For example, let's assume you license VoiceAlert and receive 5 percent of all sales of the VoiceAlert feature. At one time, you thought the company you were licensing VoiceAlert to would charge consumers $2 per month and you would receive $0.10 of that amount each month. However, the licensee has now decided to bundle your feature together with two other voicemail features for a total price of $5 per month. What amount should you receive your 5 percent on? What if the

licensee claims that your feature is only worth $0.50 per month—your 5 percent only equals 2.5 cents of that! Because these situations can and often do arise, it's important that you set some sort of expectation for the cost of your feature, such as a minimum amount you'll receive for it.

- **Continued ownership.** Although you may have exclusively licensed the patent, you still own it! Therefore, you need to state this in the agreement so it's completely clear.
- **Ownership of improvements.** Technology is constantly improving, right? What if the licensee improves your invention and files for a patent on the improvement? As you know from our earlier discussions, they would own these improvements. Therefore, you need to think about whether you want to own improvements, or at least jointly own them with your licensee. Obviously, it would be best to own them.
- **Notification of infringement.** Because the company that licenses your patent will be out in the marketplace, they may be in a good position to observe other companies infringing on your patent. If they come across any infringers, they should notify you immediately and let you know as much as possible about the perpetrators so you can take action against them.

- **Term.** For how long are you going to let the company use your patent? If the licensee is unsuccessful in marketing your voicemail feature, you should have the ability to end the relationship and engage with someone who can effectively market VoiceAlert. As the patent owner, you should have some flexibility to terminate a license agreement. Many individuals have entered into a licensing arrangement only to realize later that they received less than they could have.

When You're the One Taking a License

Why would you ever want to license another company's patent? Another company's patented invention may be of great benefit to you or your company. Perhaps, in our preceding example, you're competing against another voicemail company that has the identical features in their product as you do in yours. You could, perhaps, use the VoiceAlert feature on an exclusive basis to differentiate your product.

Be Sure You Get It All

The patent you're licensing may only be one of a number of patents owned by the licensor. Don't get trapped into licensing one of their patents, only to realize later on that you actually need to license other patents from them. Also, be sure your license covers any related patents, such as continuations or continuations-in-part or divisionals. Your licensor could also have rights to the technology in foreign countries, so you'll want to be sure if you're headed to international marketplaces that you have your rights secured. Here are some things you should consider when licensing to or from someone else:

- **Your licensor should protect you.** What happens if you've licensed a patent exclusively and you find some other company out there infringing on it? You paid for this right, so should the company you're licensing from enforce the patent against that other company? Of course! Therefore, be sure you make it a requirement that your licensor diligently pursues any infringements of the patent.
- **Release if the patent isn't good!** If, in the future, the patent you've licensed is found to be invalid or unenforceable, you should be released from having to pay the licensor any further royalties. The licensor is probably not going to reimburse you for license fees already collected, though.
- **There may be other patents you need.** When negotiating for a patent license and how much that license may cost, realize that you may need to license other patents in the future. There may be other companies with patents that are useful or necessary for you to have a good product. If you use up all the money you can spend on licensing for one patent, you won't have any room for the next license you need to take.

As we explained, this is just a quick list of items to consider when licensing to or from someone. We do recommend consulting an attorney before you enter into a licensing arrangement.

So What Is a Patent Worth?

You're probably wondering how to determine the value of a patent. Up front, you should realize it's not a perfect science, because a patent is perhaps one component of the overall value of a product. Other factors include the brand that it's marketed under, the other patented or nonpatented features included in the product, the competitors's products in the marketplace, and so on.

When you're trying to value a patent for purposes of selling that patent, there are a few methods used by valuation experts. However, if you're licensing the patent, it's another story. Just because you arrive at the overall value of a patent does not make it obvious what you should charge for a license. Armed with the valuation information, you can at least decide whether the price is in the ballpark. Also, valuation information is useful for deciding whether to charge some up-front amount (which may be a good idea if the deal is speculative) or focus entirely on a contingency based on a licensee's sales. Of course, you must remember that a contingency might amount to no revenue, and you should have the ability to terminate the agreement or at least make it nonexclusive if this happens. In any event, let's go through the various valuation methods.

Valuation Methods

There are three basic approaches used by valuation experts to value patents: *cost to replace*, *market*, and *income*. The cost-to-replace approach is the amount it would take to create a comparable patent that's as highly regarded and useful. Therefore, to figure this out, you would need to take into account all of the research and development costs spent to create the patent (assuming all the same costs would needed to be undertaken today). The market approach tries to determine how much it would cost to license the patent if someone else owned it. Of course, this method is difficult under the current circumstances, because that's precisely what we're trying to figure out! The general concept behind this approach is that value has been created from owning the patent and now you don't have to go out in the market and license it. The income approach is today's value of all the money you're making from a particular patent. This is sometimes difficult to figure out, but it can be a pretty powerful indication of value when done correctly. In the following section, we discuss some of the challenges we believe exist with current approaches to valuation of patents.

Important Valuation Considerations

Typically, intellectual property valuation firms base their findings of value on a set of assumptions that the patent is valid and enforceable and that the claims broadly cover the technology. Of course, in the real world, these assumptions are often not true (or at least not to the extent that valuation firms think). Therefore, there are many factors that need to be considered when valuing a patent. Some of these are as follows:

- What's the true scope of the patent? How broad is the patent really, and are there easy and useful work-arounds? The easier the work-arounds, the less the patent is worth.
- Is the patent only good in the United States, or are there related patents in foreign countries with large marketplaces for the invention?
- Are there any problems with the file history of the patent? Did the patent attorney make limiting statements that could narrow the claims of the patent? If so, the patent is worth less.
- Has the patent been tested in court? Companies tend to greatly value patents that have been tested in court and found valid and enforceable.
- Is the patent everything that's needed to produce the product, or will the licensee (or someone buying the patent) need other patents to produce the product? If your patent is only one small piece of the puzzle, it won't have as much value because a licensee will have to reserve some money to spend on the other necessary pieces.

Pure Profit Licensing

Many companies are beginning to realize the vast opportunity they have in licensing patents. Because a tremendous investment has already been made in developing the technology and obtaining an issued patent, the opportunities to make money off of the patent become clear. For example, IBM, Texas Instruments, and Lucent have for years licensed their patents to technology companies worldwide and have reaped billions of dollars doing so. Patent licensing is a rapidly growing part of many companies' business models. Most important, patent licensing can be very profitable because there is little if any additional cost associated with entering into a license (as we said, the investment has already been made).

The Least You Need to Know

- Just as with regular property, you can license or sell intellectual property.
- There are many elements to a patent license agreement that require consideration.
- Patent licensing can be a very profitable opportunity for a company.

Chapter 12

Online and Offline Resources

In This Chapter

- Online resources for patents
- General online legal services
- Swimming with the sharks (lawyers)

You've learned a ton about patents in this book. Before we depart, though, we want to share some information regarding online resources that can further your knowledge, as well as some tips for hiring the right patent lawyer to assist in your process.

Online Patent Resources

There are many websites that will provide you with excellent additional information regarding patents—in the United States and abroad. Of course, you can always mail or e-mail us—the authors—should you have any questions, and we'll be pleased to point you in the right direction. Also, we provide a variety of resources, including a periodic electronic newsletter you can sign up for, through our website at www.lavagroup.net.

Government Agencies

Government agencies, both domestic and foreign, can provide significant benefits to users.

The Mother of All Patent Websites

The official U.S. government website for patents (the United States Patent and Trademark Office [USPTO]) is www. uspto.gov. From this site, you will be able to access a vast array of resources. If you click the Patents link on the home page, you can access a series of categories that will help you with your patents questions, including search aids that enable you to research previously issued patents.

The USPTO also publishes a variety of resources and circulars that can be accessed through the site. These items often contain analysis and descriptions of changes to the patent requirements as well as interpretations of specific regulations. Because these resources are published by the USPTO, they have a high degree of reliability (as opposed to some outside person's interpretation of the law). If you can't find the answer to your question, you can always call the USPTO. Their telephone number is provided on their website, and they have a help desk that is dedicated to answering questions from the public.

Just the Facts

The USPTO now *encourages* applicants to do as much electronically as possible (both electronic filing and faxing). Terrorist acts such as the anthrax scare have

World Intellectual Property Organization

The World Intellectual Property Organization (www.wipo.org), also known as WIPO, is one of the specialized units of the United Nations and focuses on enforcing 23 international treaties related to intellectual property. WIPO administers the Patent Cooperation Treaty, hosts discussions among countries to further develop cooperation related to patents, and assists lesser-developed nations in their institution and modernization of patent systems (as well as other intellectual property needs).

U.S. Associations

There are associations in the United States and abroad that provide excellent reference information so you can learn more about intellectual property and obtain helpful resources. We have indicated a couple of them in the following list:

- **The American Intellectual Property Law Association (AIPLA).** The AIPLA can be found at www.aipla.org. The AIPLA is more than 100 years old and focuses on ensuring that IP lawyers maintain high ethical standards, improve legislation relating to intellectual property matters, and educate the public on these matters.

- **The Intellectual Property Section of the American Bar Association (ABA).** The ABA has a section dedicated to intellectual property. Formerly known as the Section of Patent, Trademark and Copyright Law, it is now called the Intellectual Property Section. It has been around since 1894 and focuses on developing the system for the protection of intellectual property rights. The Section has 20,485 members as of August 31, 2003, making it the world's largest intellectual property organization.

Patent Search Services

As we've described, it is often useful to conduct a patent search (called a *patentability search*) before you invest a significant amount of money into applying for a patent. Many firms provide these types of services, including Source Translation Optimization (www.bustpatents.com), Specialized Patent Services (e-mail them at patmark@interramp.com), and Express Search (www.expresssearch.com). These firms can conduct domestic or international searches on patents and regular literature.

As we mentioned earlier, a search is not an exact science, and you have to explain in great detail the type of prior art you're searching for. You can also indicate the types of resources you want reviewed, such as United States–issued patents and published applications only, foreign patents and applications, and general literature. Obviously, depending on your request, the costs may vary. A U.S. patent and application search may cost in the range of $300 to $1,000 and above. It typically takes two to three weeks to receive your results. After you receive your results, you may want to hire a patent attorney to interpret them based on your objectives (for example, trying to invalidate someone else's patent or determine whether your invention is patentable). These costs may seem significant now, but they are low compared to the challenges you may face in the absence of these results.

Private Online Resources for Understanding Legal Issues

You may have an interest in becoming a bit better versed in patent law or other legal issues. There are a variety of great websites available to you for just this purpose. We list a few here for your review:

- **FindLaw (www.findlaw.com).** FindLaw provides online legal information and solutions for the public and is the most-visited legal website. These resources include a directory of lawyers and legal professionals.
- **GigaLaw (www.gigalaw.com).** GigaLaw provides legal information for Internet and technology professionals, Internet entrepreneurs, and the lawyers who serve

them. There are a variety of references to broad legal areas and articles as well as summaries about those areas. GigaLaw is produced exclusively by lawyers and law professors.

- **Invent Now (www.inventnow.org).** Invent Now is a nonprofit organization that provides programs, content, and other information to inventors and encourages creativity and innovation. Invent Now's mission is to celebrate and foster the spirit and practice of invention.

Using a Patent Attorney or Patent Agent

We mentioned earlier that it is usually a good idea to use an attorney or agent for a variety of patent-related matters, including filing an application. That being said, there are many aspects of the application you can prepare, which will greatly reduce your expenses. The objective of this section of the book is to provide you with some guidance regarding using a patent attorney or patent agent.

Patent attorneys are specialists within the field of patent law. An attorney's experience in drafting an application or responding in an appropriate manner to an office action as well as negotiating with a patent examiner can be very helpful and ensure you obtain the broadest possible protection for your invention. When it comes to matters such as an opposition, patent infringement, or licensing of a patent, using an attorney is highly recommended. Therefore, we want to take you through some of the benefits of using an attorney in your patent matters:

- **Greater assurances.** Although the information in this book will help you a great deal with the patent process as well as introduce you to a variety of other important issues and considerations, it's always good to have someone on your side who is an expert on patent issues.
- **"Safehouse" for your records.** For patent prosecution, your attorney likely maintains organized files including the correspondence sent to and received from the patent office, which is vital if there's ever a dispute as to dates and ownership matters.
- **Docketing of your patent application and issued patent.** As you've learned, many critical deadlines arise before and after the filing of a patent application, such as the deadlines for filing a foreign application, maintenance fee deadlines, and office action response deadlines. Your patent attorney should maintain a software *docketing system* that keeps track of these dates and ensures that matters are handled in a timely manner.
- **Involvement in related matters.** Oftentimes, the process of applying for a patent or dealing with another patent issue is only one part of a much larger effort you're

undertaking (for example, patenting an invention may also be related to forming a company based on that invention). If this is the case, it is probably a good idea to have an attorney or firm who can assist you in the overall process to ensure that you fully understand your rights and obligations.

Legal-Ease

A patent **docketing system** is a software program that tells an attorney the critical dates when items are due. It also reminds the attorney to communicate with the client to ensure that decisions are made in enough time.

Picking an Attorney

Choosing the right attorney for your patent matter really depends on a variety of matters and what you believe you'll need from the attorney in the long run.

References

In every service profession there are good, okay, and downright bad professionals. With respect to intellectual property professionals, this rule holds true. Therefore, it is important that you do your homework and select an attorney who has *demonstrated* the ability to handle the work you will ask him or her to do. The best way to do this is to obtain a referral from someone you trust who has worked with the attorney in the past on a related matter. In the absence of a personal referral, you should ask the attorney for two or three references.

In addition to references, you should ask the attorney questions related to each of the other topics in this section (time, cost, docketing, experience in a related matter, and personality).

Time

One of the biggest complaints about attorneys is their responsiveness to a client's needs. Failure to handle a client's request in a timely manner is usually because of the lawyer's busy schedule. This is not such a terrible thing though—often a busy lawyer is a good lawyer. Therefore, you should reach agreement on the exact information the lawyer needs from you and the time frame for filing a patent application once the information is submitted. Then, hold him or her to it. The same applies for any other tasks you may request of the lawyer.

Just the Facts

Did you know that most states have a bar rule specifically focused on attorney response time to client requests? The state bar rules often require prompt compliance with reasonable requests for information from the client.

Experience in Related Matters

As we stated earlier, a patent issue may be part of a larger effort you are undertaking, such as launching a business. You may also have other forms of intellectual property you want to protect (such as trademarks or copyrights). If this is the case, you may need more specialized knowledge of your attorney in the other areas of intellectual property.

There are many attorneys who, either individually or through their law firm, can provide a full array of intellectual property services. If you believe a patent is the only type of intellectual property protection you will seek and you are forming a new business or have other ongoing legal needs related to your business, you should be sure the lawyer, or the firm at which he or she practices, handles this type of corporate transactional work or has a close relationship with a firm that can do this work. To ensure that the attorney handling your patent matters is qualified, you should ask the attorney specific questions regarding his or her experiences, with specific emphasis on patent issues, including whether the firm is able to docket the patent application and ensure that critical dates are not missed.

Technology Expertise

Your invention may be in a specific technical area. All patent attorneys are required to have a technical background, but those backgrounds vary greatly. You will want to be sure the patent attorney you choose has sufficient expertise in that area.

For example, if your invention is a new type of chip board for computers, you will need someone with specific electrical experience, rather than a mechanical engineer or a biologist.

Personality Fit

One item that is often overlooked is the personality fit between the client and the attorney. Many people possess preconceived notions about the personalities of attorneys. They're tough, overopinionated, nerdy, sarcastic, introverted, extroverted, or worse! In reality, there are different personalities of lawyers, and you can find one who fits your unique needs. Most important, you should feel comfortable to approach and discuss any matter with your attorney and know that he or she will respond to you in a quick and informative way.

Just the Facts

Lawyers can be a big help but have also caused clients great frustration over the years, hence the lawyer jokes. A Google search for lawyer jokes uncovered 162,000 hits!

Cost

The most important item to a client is typically the cost of an attorney. It is also likely the reason that many individuals either don't file a patent application or attempt to prepare one themselves. Using a patent attorney for your patent applications can be an expensive process, costing many thousands of dollars. However, rather than making a decision solely based on cost, please be sure you consider the other factors and benefits described earlier.

Patent efforts are typically broken down into a variety of actions, such as conducting a patentability search, reviewing the search results, preparing and filing an application, and responding to communications from the USPTO. That being said, you may want to request a flat fee quote for the application process (through filing) and the follow-up communications and compare that to other providers. Our research shows that attorneys charge from $3,500 to $12,000 and more to file an application and $1,500 to $2,500 for each follow-up effort. If issues arise during the prosecution process, costs can rise substantially.

The Least You Need to Know

- There are many great online resources to increase your knowledge about patents and keep you up to date on any changes.
- You can always call the USPTO for the answer to a specific question. The USPTO has a general help desk that is available to assist individual inventors.
- It is often helpful to use an attorney or patent agent with your patent matters.

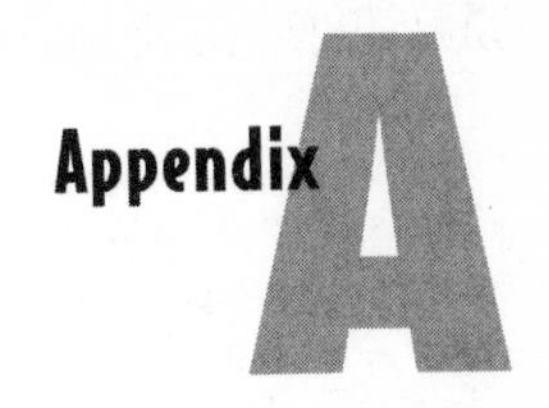

Further Reading

Department of Commerce, The. *Patents: A Practical Handbook*. New York: Dover Publications, 2000.

Docie, Ronald Louis, Sr. *The Inventor's Bible: How to Market and License Your Brilliant Ideas*. California: Ten Speed Press, 2004.

Elias, Stephen and Richard Stim. *Patent, Copyright and Trademark: An Intellectual Property Desk Reference*. California: Nolo Press, 2003.

Hitchcock, David. *Patent Searching Made Easy: How to Do Patent Searching on the Internet and in the Library*. 2nd Ed. California: Nolo Press, 2000.

Isenberg, Doug. *GigaLaw Guide to Internet Law*. New York: Random House Trade Paperbacks, 2002.

Levy, Richard C. *The Complete Idiot's Guide to Cashing In on Your Inventions*. Indianapolis: Alpha Books, 2001.

Pressman, David and Jack Lo. *How to Make Patent Drawings Yourself: Prepare Formal Drawings Required by the U.S. Patent Office, 2nd Ed.* California: Nolo Press, 1999.

Pressman, David and Richard Stim. *Nolo's Patents for Beginners, 3rd Ed.* California: Nolo Press, 2002.

Glossary

business method patent A utility patent that includes claims seeking to protect a new, useful, and nonobvious approach to pursuing a particular solution for a business challenge (for example, Amazon.com's one-click ordering system, which enables customers to quickly order items from).

claim The part of a patent document that defines a protected invention by listing the elements and limitations of the protected invention.

claim the benefit of *See* claim the priority date of.

claim the priority date of Identifies the legally recognized filing date for a patent application. In certain circumstances, you can claim the filing date of a prior application as the filing date for your application. This is a claim to the priority date.

conventional patent A type of patent used to protect a new and useful process, machine, manufacture, or composition of matter, or an improvement to any one of these items. This is synonymous with utility or nonprovisional patents.

copyright A way to protect the expression of an idea—usually referred to as a "work"—that covers a variety of items including literary works (books), sound recordings (music), visual arts (statues), performing arts (movies), and serials and periodicals (newspapers).

declaration A document that must be filed with a patent application or shortly after the application is filed and must be signed by each inventor, providing address information for the inventors and sworn statements as to inventorship.

dependent claim A claim that modifies or adds additional limitations to another claim.

design patent A type of patent that is used to protect the ornamental aspects of a device and prevents others from copying the visual characteristics of an article such as the shape, the ornamentation applied to the surface of an item, or both.

docket number A reference label that is selected by the person, attorney, or law firm filing a patent application and is used to identify the application.

examiner An employee at the USPTO who is tasked with examining the merits and allowability of a patent application.

file history The collection of correspondence that goes back and forth between the USPTO and the applicants or the applicants' attorneys.

filing receipt A notification from the USPTO indicating the date, serial number, and other particularities of a received patent application and serves as formal proof and verification that you filed the patent application.

independent claim A claim that stands by itself and fully defines all of the limitations of a patent-protected invention.

invention disclosure document A filing with the USPTO that can be used to establish the date on which you invented something.

maintenance fee The fees that must be periodically paid for an issued patent.

multiple dependent claim A dependent claim that modifies more than one additional claim.

nondisclosure agreement A written agreement between parties under which a first party agrees to disclose information to a second party and the second party agrees to not tell anyone else.

nonprovisional patent A type of patent that is used to protect a new and useful process, machine, manufacture, or composition of matter, or an improvement to any one of these items. This is synonymous with utility or conventional patents.

notice of allowance An official communication from the USPTO indicating that claims are allowable.

notice to file missing parts An official communication from the USPTO indicating that the filing package was missing items such as the filing fee, declaration, proper transmittal sheet, and so on.

office action Refers to an official communication from the USPTO that provides the USPTO's opinion regarding the patentability of the filed claims. The office action can include rejections, objections, or indications of allowable claims.

Paris Convention Was signed into effect in 1883 and a revised version was introduced on July 14, 1967. A primary purpose of the Paris Convention is that it obligates member nations to provide substantive protection for the procurement, maintenance, and enforcement of industrial property.

patent A monopoly in an invention, granted by the government, that allows an inventor to prevent others from making, using, selling, offering to sell, or importing products or services that incorporate or require the invention.

Patent Cooperation Treaty (PCT) An international agreement that facilitates obtaining international patents by establishing entities that can service as central depositories for filing patent applications that can ultimately be filed in over 100 countries that are members of the PCT.

plant patent A type of patent that is used to protect any new variety of plant that you invent or discover and can asexually reproduce, other than a tuber-propagated plant or a plant found in an uncultivated state.

prior art Information related to your patent filing that demonstrates someone else invented your claimed invention before you did. This information may be used to prove that your invention is not new or is obvious.

priority date The earliest date that a patent claims as the filing date. The priority date can be the filing date of a related foreign patent or an earlier filed U.S. patent application.

process patent A utility patent that includes claims seeking to protect new, useful, and non-obvious series of actions, changes, or functions bringing about a desired result or solution to a challenge—such as business methods or software systems/applications.

prosecution The negotiation of the patent application between the inventors (or attorney representing the inventors) and the USPTO examiner assigned to the application.

prosecution history *See* file history.

provisional application for patent A procedure for filing an invention description with the USPTO and obtaining a priority date or filing date. A provisional application for patent is not a patent application and does not get examined by the USPTO.

restriction requirement An official communication from the USPTO indicating that the examiner has determined that the claims of a patent application are directed towards more than one invention and that the applicant should select just one set of claims to prosecute.

return receipt postcard A postcard that lists a description of each item that is included in a package filed with the USPTO and is self-addressed and includes postage. The return receipt postcard serves as informal confirmation that the USPTO received your documents.

shop rights doctrine An area of law that allows a company to obtain a license in a patent if the invention was developed as the result of the company providing the resources, the lab area, the office space, and so on, to make the invention.

specification The part of a patent or patent application that includes the title, summary, background, and detailed description of the invention.

statutory invention registrations A procedure to file a document with the USPTO that allows you to publicly disclose an invention so that no one can subsequently claim ownership.

trademark Words and symbols (called *marks*) that help consumers identify the origin of certain goods and act as the brands a company may use to offer its products or services.

utility patent A type of patent that is used to protect a new and useful process, machine, manufacture, or composition of matter, or an improvement to any one of these items. This is synonymous with nonprovisional or conventional patents.

World Intellectual Property Organization (WIPO) An administrative arm of the United Nations that supports international trade and promotes reciprocity of IP recognition and protection among various nations.

Index

A

B

C

D

E

F

G–H

I

J-L

M

N

O

P-Q